volcano

MAURICE AND KATIA KRAFFT

Introduction by EUGÈNE IONESCO

Poems by MAX GÉRARD

Translated by JOHN SHEPLEY

volcano

Harry N. Abrams, Inc., Publishers, New York

FOR OUR PARENTS

Library of Congress Cataloging in Publication Data

Krafft, Maurice.
 Volcano.

 1. Volcanoes. I. Krafft, Katia, joint author.
II. Title.
QE522.K713   551.2'1   75-16298
ISBN 0-8109-0557-4

Library of Congress Catalogue Card Number: 75-16298
Copyright © 1975 by Draeger, Imprimeurs, Paris
Published in 1975 by Harry N. Abrams, Incorporated, New York
Printed and bound in France

CONTENTS

INTRODUCTION

A doctor specializing in natural medicine, obviously a Swiss, once advised me to live according to nature's rules. It was the second or third time I had gone to Switzerland, to a place between Basel and Zurich. I saw nothing but hills. I had almost come to believe that high mountains were only an invention of the boasting Helvetians. My geography teachers must have lied, I thought, since I saw nothing but hills, green meadows, picturesque towns, lakes most likely set in place by cinema scene painters. It was only on my third or fourth trip that I was able to see, and later approach, the high, barren, foreboding, dangerous, sublime mountains. But at first even the forests where I took walks were furrowed by wide roads for automobiles. Here and there were also benches, for the repose of intrepid walkers. I was able to tell myself: How gentle and civilized nature is! Where else but in cities do you find the jungle? There were also orchards and vegetable gardens. I saw worthy Swiss cows passing tranquilly by. To be happy and serene all you had to do then was to identify with one of these countless well-kept trees, with these clean and healthy cows. Man's unhappiness comes from the fact that he lives with trepidation in cities—amid noise, tainted air, the sourness of desire, violence. Everything is so calm and beautiful in a garden. Yes, everything is so beautiful, so calm, in a garden!

"Not at all!" the gardener tells me. "At all times, for hours each day, I have to maintain order. I must keep the so-called weeds from destroying the good grass. Each plant wants to encroach on the domain of others. I destroy to prevent general destruction: I clear away the mistletoe and ivy that feed on the sap of the trees. I must kill the caterpillars. This beautiful and serene garden, which you see as so harmonious, is a field of battle and carnage where I am the tyrant, the gendarme, the killer. Each tree, each plant, each blade of grass is in danger, and each is a danger to the other citizens of the vegetable world. My job is repression."

Indeed, you wonder how Rousseau's conception of nature and the noble savage, in its naïveté, could have lasted so long. One need only look at a drop of water under a microscope to see the relentless battles that go on there and in the cells of our bodies: organisms spend their time gobbling each other up. Insects strike and devour each other. No creature is "nice." In both the microcosmic and macrocosmic world, objects fight as though they were alive: they attract and repel each other, and universal equilibrium results from the laws of gravitation, from the dynamic equilibrium of opposing forces. I am waiting, from one moment to the next, for this precarious equilibrium to go to pieces and for the world to be turned upside down or become a glutinous mass. If matter exists, the whole universe can be reduced to a speck of dust. If matter does not exist, as certain scientific men think, there is nothing save material energy to give the illusion of substance. How can we hope for peace, how can we think to found a world on nonviolence, when violence is existence itself? Nonviolence is a paradox that no longer holds. Everything is aggression, conflict. One cannot be at peace with anyone, since one is not at peace with oneself. We struggle constantly against ourselves, and we never know which part of ourselves will be stronger than the other. If I stop fighting, I fall into depression, I die. My corpse will put up some resistance to the natural agents that will decompose it. The war will go on. For the moment, I exist. Passions slumber in me that may explode, then be held in check again. Jets of rage or joy lie within me, ready to burst and catch fire. In myself I am energy, fire, lava. I am a volcano. Most often, I am half asleep: my craters wait for this continual boiling to rise, emerge, satisfy its instincts; for my incandescent passions to pour out, ignite, and spread forth in an assault on the world. So long as I do not become an extinct volcano, a worn-out volcano.

If one wants to live in conformity with nature, one must take as a model floods, thunder and lightning, tempests and cyclones, volcanic eruptions. I said just now that I was expecting a cataclysm at any moment. The final apocalypse lies in wait for the world, which is approaching its end. But the apocalypse is also daily; we live it while waiting for the big one. Let us be apocalyptic or let us be nothing; let us be violent on condition that our violence be guileless, that our explosions be candid, that our battle take place in joy. The universe is volcanic—let us consciously be volcanoes.

I happen to have premonitions. A long time ago I found myself in a Balkan town. All of a sudden I wake in the middle of the night, screaming: "Earthquake!" There was no earthquake. My wife tells me I'm having an awful nightmare, I lie down and go back to sleep. The next night, at the same hour, I again wake with a start and scream: "Earthquake!" My wife tells me I'm crazy and to go back to sleep. Hardly has she uttered these words when we hear horrible cracking sounds and the house begins to pitch like a boat in a storm. Pictures come off their hooks, plates fall and break. We rush toward the stairs, but then it all stops. I look back as with nostalgia on those moments of terror that I would not want, and yet do want, to relive. The advantage of terror is that it makes boredom impossible. Neurasthenia lies in wait for peaceful souls. Our house had stood up, with only a few cracks in the walls, but many buildings in the town had collapsed. We learned that twenty-four hours before the seism, the cows had bellowed in their stalls and had pulled on their ropes in an effort to flee. The horses, neighing and stamping, had broken the doors of the stables. Dogs had begun howling. I was the only man, along with the dogs, cows, and horses, to have had a premonition. I was very proud of it.

Some years later we found ourselves in the Cyclades, on Thíra, or Santoríni. The shape of the island is strange, nearly circular—a mountain like a ring in the sea. In the middle, a magnificent volcano. It seems that Hephaestus, being displeased one day, had taken the island of Thíra in his hand and thrown it some distance, like a stone. It had fallen in the sea not far from Italy, giving birth to the volcanic island of Stromboli. But in uprooting the center of the mountainous island, Hephaestus had left its edges, with the volcano in the middle. They say that if one were to put Stromboli back in its former place, it would take up precisely that part of the island that was pulled up.

I had never seen stones and pebbles so beautiful as those I was able to see on Thíra: red, green, chestnut. The little town was at the top of the mountain; you went there by donkey. The spectacle was breathtaking. From the heights you could see the nearly circular form of the island, at its center the giant—the threatening volcano. If you turned your back on the volcano, you saw a gentle slope descending peacefully toward the sea, like a large garden full of flowers, tomatoes, vines. To dispel our moments of anxiety, we would look at that tranquil soil. But in moments when the spirit was ready for the confrontation, it was toward the volcano that we looked.

We had been there for a few days when one morning I was seized with terror. "There's going to be an eruption immediately. Let's hurry up and leave," I said to my wife. Since she now believed in my premonitions, we left the island three days later, for the boat came only twice a week. My presentiments were justified. There was a gigantic volcanic eruption. Lava and stones shot up very high in the air, fell back on the town, and destroyed it, though it was at the summit. True, this happened nineteen years and three months after our departure. There was too much anticipation in my premonition.

I have failed to be shot. In planes, I was twice caught in storms and came close to catastrophe. A hurricane almost capsized our boat. I once had an attack of vertigo on a rock. Perched on the branches of a cherry tree, I have had to fight off a swarm of wasps. I have escaped from a railway disaster. In a stalled cable car, we had to wait three hours for help. One day a tree fell right at my feet. Some years later, another fell whose leaves grazed my back. And a few years ago I penetrated at least fifteen meters into the virgin forest.

I have not known volcanoes. I have tried to climb Salvatari, but the lava was still too hot for my tourist's feet, clad in shoes good only for the pedestrian pavements of bourgeois citydwellers.

I have seen, at first in painting, Fuji san, or Fujiyama; later, from afar, I was able to contemplate it at a respectful distance from a roadside, through the branches of a flowering Japanese tree; likewise from afar, I have seen Etna; at a little closer range, Vesuvius, but I have drunk "Lacrima Christi," the wine that comes from the vineyards cultivated on the slopes of the volcano; in a boat, I have touched the shore of dangerous Stromboli, permanently simmering and spitting flames and despite this, or just because of it, covered with white houses that squeeze together at its foot or creep halfway up the distance separating its base from its summit of boiling lava, like toreros hoping to arouse the anger of a gigantic bull with fuming nostrils while at the same time wearing it away; I have looked at the sad, humble, extinct volcanoes of the Auvergne; I have not even glimpsed Krakatau, which cheerfully destroyed half the island of the same name; nor the massif of Kilimanjaro, evoked by Hemingway in one of his works; nor Mount Pelée, nor the Icelandic Surtsey, nor the other volcanoes of Iceland; nor Paricutín, born in 1943, which has not stopped growing, nor its elder brothers in Mexico; nor those in Tierra del Fuego, Antarctica, New Zealand, the Philippines, the Antilles, the Azores, the Canary Islands, the Sunda Isles, East Africa, the Indian Ocean; nor especially the Congolese Nyiragongo, the forbidden volcano.

Hephaestus—Vulcan to the Latins—goes on forging his weapons. He avenges himself on Heaven because Zeus threw him down from the heights of Olympus. It was in falling that he broke his leg. He descended into the bowels of the earth—he is the god of fire. Sickly at his birth and ugly, his weakness has been transformed into strength, his spirit has made him beautiful, his rancor has made him redoubtable, the fear he inspires has made him superb, magnificent. Brilliant as the sun, he is Phoebus' rival. He is known in all mythologies; he is Agni, the fire god of the Vedic hymns; he is Regin, the armorer craftsman of Hialprek, king of Denmark; he is perhaps Satan. He is the patron of blacksmiths, of weapons makers; his courtiers are, in all probability, the traffickers in arms and munitions. It is he who inspired the atomic bomb. He is still alive, hidden in the depths; he sneezes in his sleep and the earth's crust trembles; he smokes; he turns over and mountains spring up. The volcanoes are his sons, he has hundreds of them. The craters, jaws of enormous cannons, are their half-opened lips, their gaping mouths, their gullets. Aphrodite mates with him in his cave. Hephaestus is still forging weapons, and other sons will be born; when there are a thousand of them, and when the sleeping ones awaken too, he will rise up with them for the conquest of Heaven; incandescent lava will cover our swamps, our reeking cities, our fields, our flowering hills; it will destroy the contours of that soil we persist in calling "ours." We are ants, but we will not die without having witnessed the most grandiose of conflagrations, without having glimpsed, spreading from thousands of giant jaws, the Explosion that will encompass the Universe.

When the earth opens…

THE ERUPTION ON HEIMAEY 1973

Between January 21, 1973, at 10 P.M. and the next day at noon, the needles of the seismographs at Hafursey and Hestfjall in southern Iceland zigzag on the register sheet and indicate earthquakes. A small seismic crisis has just taken place—where is its focus located? Two possibilities occur to the geophysicists: either the east of the Hekla volcano, last active in 1970, or the island of Heimaey in the Vestmann archipelago, to the south of the Icelandic mainland. The second location is dismissed, since the last eruption on the island happened more than five thousand years ago when Helgafell had awakened; besides, none of the 5,300 inhabitants of the island has felt the earthquake. To be sure, there is Surtsey, the volcano born in the sea in 1963, but it lies 20 kilometers southwest of Heimaey. The depth of the seisms is determined: 18 kilometers underground. No one worries about it, the earth trembles so often in Iceland. The seismographs subside, but not for long; on January 22 at 10:30 P.M., the seisms begin again, more numerous this time and very close to each other—their focus is at a depth of only 2 kilometers! What is happening and where? The answer is not long in coming: between midnight and 1:35 A.M. on January 23, the earth shakes three times on the island of Heimaey. Being so used to this kind of thing, the inhabitants scarcely notice it and go on sleeping peacefully. But two men, two fishermen, are walking in the streets in the eastern part of the town of Vestmannaeyjar. At 1:55 A.M., they suddenly see a line of fire behind the houses; it spreads, noiselessly and rapidly, from south to north. They have been drinking a little, of course. Convinced they are having hallucinations, they go closer to the phenomenon, and then they understand that the line of fire is a gaping fissure that spits a curtain of molten lava more than 100 meters high. A new volcano is being formed! The two men forget to give the alarm; they run home, wake their wives and children, then rush to the harbor. A few seconds later the telephone rings at the police station of the town; the incredible news spreads. The municipal authorities are immediately alerted; the police, blowing their sirens, rush through the streets of the sleeping town; the whole population awakens, jumps out of bed, and looks with astonishment at a growling, reddening volcano. The civil-defense office of Reykjavík, 120 kilometers away, is informed

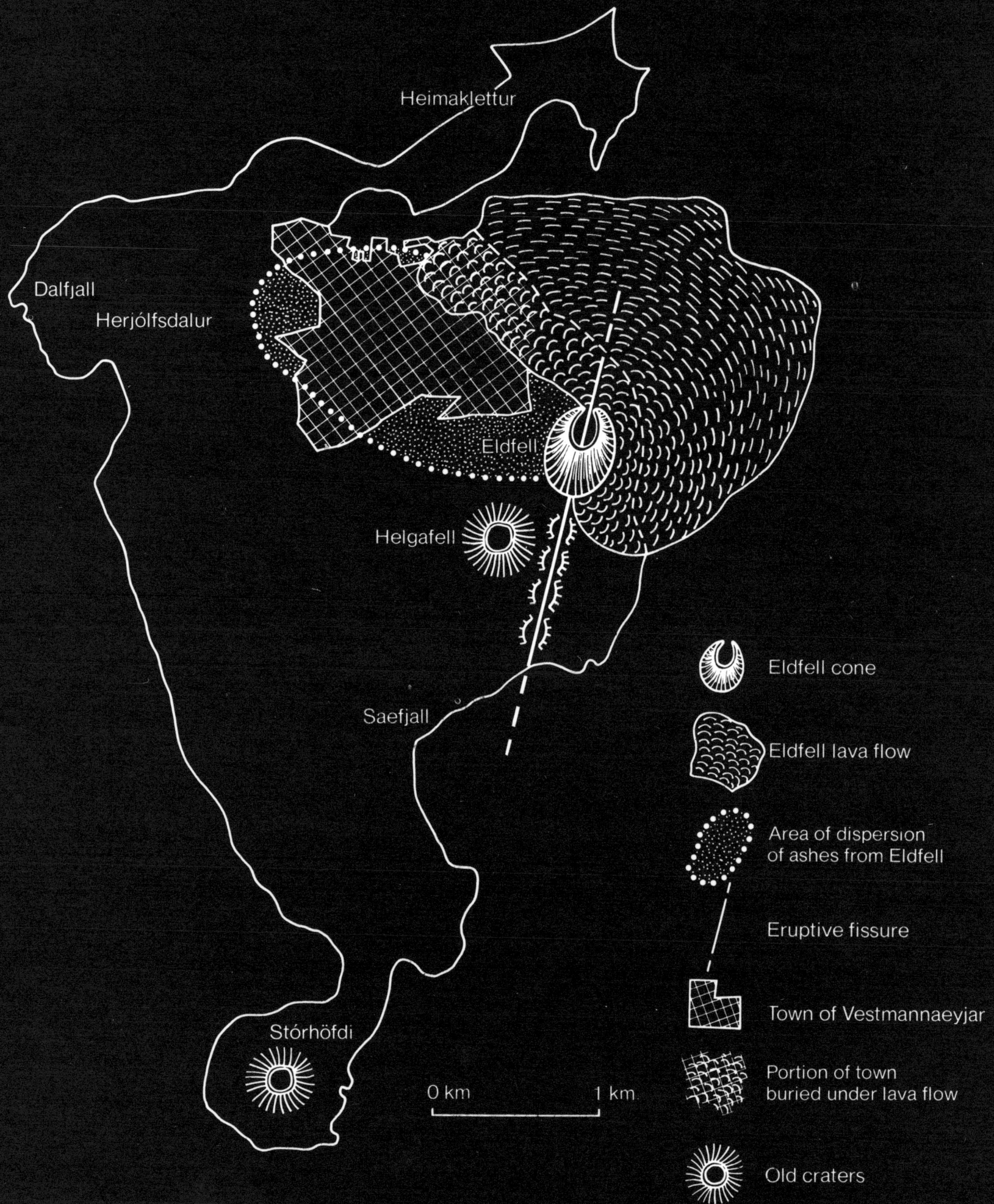

Heimaklettur

Dalfjall

Herjólfsdalur

Eldfell

Helgafell

Saefjall

Stórhöfdi

Eldfell cone

Eldfell lava flow

Area of dispersion
of ashes from Eldfell

Eruptive fissure

Town of Vestmannaeyjar

Portion of town
buried under lava flow

0 km   1 km

Old craters

1 MAP OF HEIMAEY (VESTMANNAEYJAR, APRIL 1973)

of the eruption at 2:05 A.M., hardly ten minutes after the beginning of activity; it receives the following report: "At 1:55 A.M. this day, January 23, a fissure 1,800 meters long, 3 meters wide, direction northeast-southwest, has opened on the eastern side of Helgafell, near the Kirkjubaer farm, 250 meters from the eastern end of the town of Vestmannaeyjar; it is spewing about twenty fountains of molten lava, more than 100 meters high" (figure 1). The civil-defense department, the airport of Keflavík, the hospitals, and the police are mobilized without delay. The evacuation plan for the island, worked out a few years earlier following the eruption of the Surtsey volcano offshore from Heimaey, is taken out of the files. The civil-defense authorities of Reykjavík hold an emergency meeting. Their decision is immediate: "Total evacuation of the island."

Meanwhile, in Heimaey, police cars with loudspeakers cruise the town, giving the order to all inhabitants to proceed immediately to the harbor, carrying only what they need to protect themselves against the cold. Fortunately, that night the weather is good and the sea calm. Luckily too, the seventy-seven fishing boats of Heimaey are there, having taken refuge in the port the day before because of a very rough sea. Moreover, they are ready to leave, in perfect condition and with their tanks fueled, since it is the beginning of the fishing season. One may well wonder how the evacuation would have been possible without this combination of fortunate circumstances. The victims embark at once—with no panic, in orderly fashion, in the middle of the night. The order is given: all boats are to proceed to the port of Thorlákshöfn, four hours away by sea to the north, on the southern coast of the Icelandic mainland. The civil-defense authorities of Reykjavík are already requisitioning all the buses of the capital; they are able to drive speedily toward Thorlákshöfn because the road has been closed to all other traffic. All the restaurants of the port are told to make soup to welcome the unfortunate fishermen and their families.

The first plane lands on Heimaey one hour after the beginning of the eruption, and it is soon followed by American army helicopters stationed at Keflavík. Old people and hospital patients are evacuated in three hours. All planes able to fly from southern Iceland will land on the Heimaey runway that morning. Around 8 A.M., six hours after the beginning of the eruption, the last inhabitants are evacuated; approximately 5,000 people have been transported to the mainland during the night. Remaining on the island are 150 men, mostly policemen and municipal employees. This evacuation is an achievement, for if one takes into account the total number of inhabitants in Iceland and their logistic means, it would be the equivalent in France, for example, of unexpectedly transporting one million people in the middle of the night.

That Tuesday morning, January 23, most of the people of Reykjavík get out of bed without knowing anything of the catastrophe. They learn about it from the radio. It is hard to believe that, just the day before, Heimaey farmers were working on the very spot where the fissure opened: there had not been a fumarole, no sign of a crack, only good fertile soil.

Already long columns of vehicles loaded with the refugees are arriving in Reykjavík; the schools are taken over, and six thousand mattresses installed. Appeals are made to the population to shelter their compatriots. The civic spirit of the Icelanders is equal to the occasion, and by that very evening most of the inhabitants of Heimaey have been lodged with families in the capital.

Hardly has the evacuation of human beings been carried out than that of the cattle, material belongings, and the eight hundred cars on the island begins. Only cross-country vehicles, for which their owners are reimbursed, are left behind. Those houses most threatened by the new volcano are emptied of their contents. Furniture, electrical appliances, television sets pile up on the wharves of the port, waiting to be transported to Reykjavík. The stock from department stores is transferred to trucks and loaded on cargo ships. Food supplies are requisitioned for volunteers remaining on the island.

On January 24, only three fountains of lava are still quite active over a length of 300 meters, in the northeast part of the fissure. Volcanic activity intensifies; streams of basalt lava are spewed forth by the fissure, flow east, north, and northwest, and reach the sea. Their contact with the cold water of the ocean provokes an imposing release of steam, which merges with volcanic gases to produce a plume rising to a height of 8 kilometers.

Incandescent volcanic bombs fall on the Kirkjubaer farm, setting it afire. A cone of ashes forms on the fissure; it grows and rapidly reaches a height of 100 meters, since the vents are spitting forth an average of 100 cubic meters of molten rock per second.

Faced with the scope of the cataclysm, the authorities decide to save the supply of frozen fish stored in the port installations, which represents some millions of dollars, since Vestmannaeyjar, being the largest fishing port in Iceland, handles more than 20 percent of the country's fish. Night and day, fish is loaded on ships. Likewise, the important municipal documents, money from the banks, and the contents of libraries are removed.

On the twenty-fifth, weather conditions deteriorate and a violent east wind blows ashes over the town, which is quickly covered by 10 centimeters of ashes; it is the beginning of the martyrdom of this northern Pompeii. A thick stream of basalt advances the length of the mole that protects the port from the waves of the open sea. Volcanologists arriving on the scene keep watch, while American satellites regularly make infrared photographs of the hot points of the island. The fall of incandescent blocks becomes more momentous, and the following night nine houses burn in the eastern part of the city. On the twenty-sixth, the fall of ashes reaches an accumulation of 1 meter an hour near the volcano. Windows are broken by volcanic bombs and by the shock wave of explosions; ten new houses succumb to the flames. The groundwater in the pits mounts 10 meters, and its temperature rises 10°C. The chaotic stream of lava has now advanced 500 meters into the sea. The next day the wind blows from the west and carries the ashes toward the ocean, away from the town; everyone hopes this will last, but twenty-four hours later volcanic dust again falls on the town. The volcano has already emitted fifteen million cubic meters of lava; volcanic bombs fall as far as the center of the town. On the twenty-eighth the authorities estimate that sixty-two houses have been destroyed, most of them crushed by ashes, some twenty of them burned.

The following day the Althing, the Icelandic parliament, sets up a committee of seven persons to direct the salvage of Heimaey. A loan of about $6,000,000 is granted by the Central Bank of Iceland to provide for the immediate needs of the refugees. American army specialists are sent to the island to study the possibility of bombarding the east flank of the volcano so that all the lava streams flow toward the sea, away from the town, but in view of the dangers that this represents should the west flank be likewise struck, the project is given up. The only broad preventive measure consists in nailing twelve thousand sheets of corrugated iron over the windows of the houses to keep the burning ashes from penetrating the inside. Men clear dozens of roofs of their thick bed of ashes to prevent them from collapsing under the weight of the volcanic ejecta. A television camera is installed, facing the volcano, on the summit of the Heimaklettur cliffs; all day long it films the eruption, which is then transmitted on the national television channel; thus at all times the Icelanders can observe the raging of the volcano.

On January 30 the summit of the volcanic cone is 185 meters high, and nearly 110 houses have already disappeared under two million cubic meters of ashes. The thick stream of lava advances inexorably in the ocean, thus enlarging the island by more than a square kilometer.

On February 12 the lava flow threatens to close the entrance of the port, the passage being no more than 30 meters wide. All ships take to the open sea for fear of being holed up in the port installations. The lava cuts the undersea electric cables and one of the pipes bringing fresh water from the mainland. Electrical generators are immediately installed on the island. The authorities and volcanologists then decide to combat the lava, which in certain places is progressing at 6 meters an hour. Barriers of ashes 3 to 4 meters high are erected with bulldozers in front of one arm of the flow, but the experiment only proves how quickly the lava overcomes the obstacle. The only remaining solution is to pour seawater on the molten rock to cool and thus congeal it. The first attempt has positive results, since pouring 20 tons of seawater an hour on the molten rock seems to arrest its progress. The Icelandic parliament decides to increase immediately the sales taxes by 2 percent, private-property taxes by 30 percent, municipal taxes by 10 percent—this is in effect for the whole country, in order to cover the increasing deficit due to the eruption.

On February 20 five million cubic meters of ashes forming the west wall of the cone collapse toward the town, engulfing several houses. Toward the end of the month of February, the lava has covered 3 square kilometers, and the volume emitted is estimated at more than one hundred million cubic meters. The explosive activity of the crater becomes more moderate, but on the other hand the flows take on increasing importance. In the face of the catastrophe that the closing of the port would represent, the ship *Sandey* is dispatched to the scene, and with the aid of a powerful fire pump it pours 12,000 tons of seawater an hour on the stream of lava. The Icelandic government contributes about $4,000,000 to finance these combat operations against the forces of the volcano. Many countries offer Iceland their assistance, and funds are raised all over the world.

On March 25 and 26 a lava lake rises in the crater, then overflows. A stream of lava forms and flows to the northwest, straight for the town. Blindly and inexorably, the molten rock advances on the elegant houses on the east side of Heimaey. By March 28 sixty dwellings, the swimming pool, two hotels, and some administrative buildings have been swallowed up, crushed by the lava; the output, however, is meager: 10 cubic meters per second. Near the lava flow stands a hospital, completely new and well equipped: it was to have been opened to patients at the beginning of February! By the end of the month the lava flow is progressing into the port installations and destroys two of the five freezing plants that are among the largest and best equipped in Iceland; a third is crushed by the molten rock, but the machinery has been evacuated in time. The situation becomes dramatic when the flow threatens to fill the bay of the port, thus dooming forever the economy of Vestmannaeyjar, which is based entirely on fishing. Iceland then appeals to the United States. It asks for and obtains forty-seven powerful pumps, which are brought by cargo plane and immediately installed on the wharves. These pumps have a total output of 4,500 tons of water an hour. Three large pipes, 30 centimeters in diameter and several hundred meters long, are used to convey seawater from the harbor to the terrible lava flow, and the incredible feat begins. Every hour 4,500 tons of seawater are discharged against the lava front, now advancing at a rate of 3 to 8 meters a day; this has the effect of lowering the temperature of the molten rock from 1,000° to 800°C. A volume of 20,000 cubic meters of molten rock is thus cooled by 200°C and accordingly solidified every hour. As soon as the front of the flow is stabilized, bulldozers—protected against volcanic bombs by shields of steel plates—mount the lava, open passages, and level the surface to pull the pipes farther, more to the interior of the flow, in order to cool other hot points. The cooling has little effect during the first twenty-four hours, and a great release of steam is produced, but little by little, after two weeks of continuous watering, the hot points are cooled to a temperature below 100°C. Simultaneously, trucks and a steam roller erect imposing barriers of ashes in front of the lava to stop the progress of the river of fire. The operation is a success, since behind the wall of solidified lava, nearly 40 meters high, the molten rock is arrested and diverted. The port is saved. This is the first time in the history of humanity that man has been able to master a small part of the destructive force of a volcano.

The balance sheet, however, is heavy: three hundred houses have disappeared under the lava, more than a hundred are buried under ashes or burned. Damages are estimated at more than $50,000,000, at a time when Iceland's gross national product is approximately $600,000,000 a year. One third of the constructions in the town are nothing but the buried ruins of a new Pompeii. The inhabitants do not consider the island lost and set up a fish-processing plant on the western side of the town. However, they cannot yet resettle in Heimaey, for there is another danger awaiting them: gases. In all the low parts of the town, in the cellars, there is more than 90 percent of carbon dioxide, 0.1 percent of carbon monoxide, hydrogen sulfide, and methane. Volunteers accordingly install bellows in the basements of the houses they occupy. Nonetheless one man dies of asphyxiation from the toxic gases.

Several weeks later the explosive activity of the crater becomes intermittent, the flow of lava congeals. The island sinks slightly, since a great void now exists underground—where all the magma had been that has emerged from the erupting fissure; the walls of buildings crack. On July 3, 1973, Professor Thorbjörn Sigurgeirsson, who has directed the operations for arresting the lava flows, declares the eruption finished. The volcanic activity has lasted six months.

Ten days later an official name is given to the new volcano; it is to be called Eldfell, "the mountain of fire." Eldfell is 225 meters high, like its neighbor Helgafell, which was born five thousand years earlier. The new volcano has emitted 250 million cubic meters of volcanic products, of which 90 percent are in the form of basalt flows. Heimaey, which had an area of 12 square kilometers before the eruption, now has 14.5 square kilometers. The island has been enlarged, and the returning inhabitants say, "So much the better, since the port is more protected thanks to the new lava flow." By July 1973, thirteen families have already come back. Millions of grass seeds are sown by airplane on Eldfell and fixed on the ground by the use of old fishing nets. It is necessary to cover the volcano quickly with green grass in order to forget the catastrophe. The houses are cleared of their volcanic ashes. The ashes are not lacking in importance, since they are an excellent construction material, and the inhabitants put them to use in extending the runway of the airport and building new houses in the western part of Heimaey. By July 1974, four thousand people have resettled on the island, the houses are repainted, the gardens are full of flowers. A great celebration is organized to take place at the beginning of the month of August, as was done every year before the eruption of Eldfell. A huge camp of tents is set up in the valley of Herjólfsdalur, an ancient crater situated west of the town; there, amidst the general euphoria, men will light large pyres, and the valley will blaze from east to west to recall the pagan cult of the god of fire!

Vesuvius
Engraving by A. Kricher, from *Mundi Subterranei*
(Amsterdam, 1678)

TYPUS MONTIS
VESUVII
Prout ab Authore
A 1632 visus fuit

Porticı

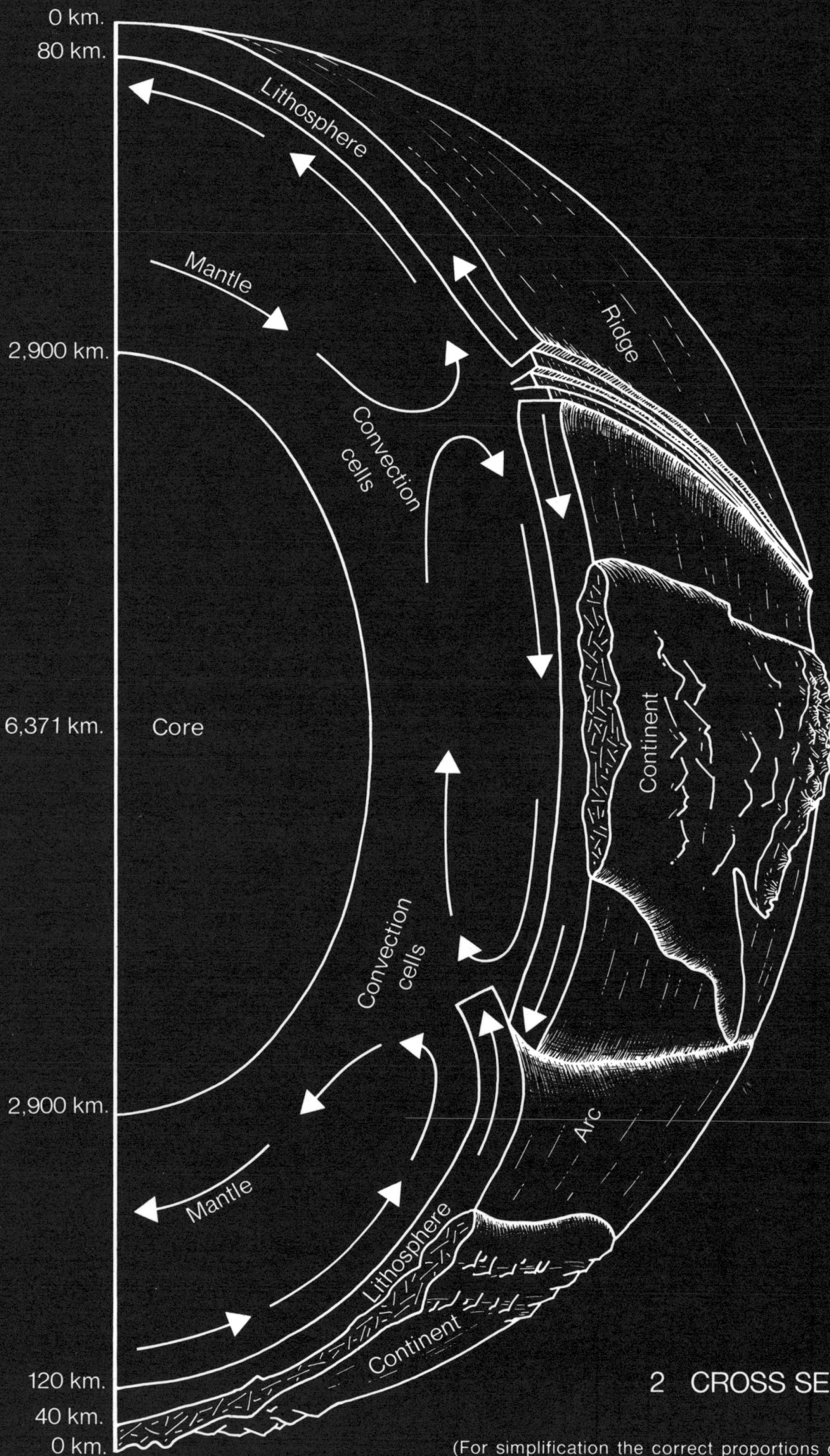

0 km.
80 km.

Lithosphere

Mantle

2,900 km.

Convection cells

Ridge

6,371 km.   Core

Continent

Convection cells

2,900 km.

Arc

Mantle

Lithosphere

Continent

120 km.

40 km.

0 km.

2   CROSS SECTION OF THE EARTH

(For simplification the correct proportions of thicknesses have not been followed

THE STRUCTURE OF THE EARTH

More than a million times a year our planet quivers, trembles, and shakes. Seismologists, on the lookout for the slightest indication that would allow them to extract a secret from the mystery of the interior of our globe, patiently sound the earth day and night with their seismographs, like doctors armed with stethoscopes. Through perseverance, genius, as well as chance, the modern geophysicist gives us an image of the structure of the earth that is doubtless more exact than the one we had at the beginning of the century.

The earth, 4.5 billion years old, is often compared to a peach—with its stone, pulp, and skin—since it consists of three concentric envelopes with well-defined characteristics (figure 2).

The earth's core, with a radius of 3,471 kilometers, occupies the central part of the globe; its outer limit stops at a depth of 2,900 kilometers. It is almost solid, consisting of nickel and iron, and is subject to pressures from 1 to 3 million atmospheres and to temperatures from 2,000° to 4,000°C. In its outer, more fluid part, the iron is in slow movement, and its circulation gives rise to the earth's magnetism in the manner of a giant dynamo.

The core is enveloped by the mantle, which begins at a depth of 2,900 kilometers and stops between 80 and 120 kilometers underground. This mantle is composed of peridotite, a rock rich in silicates of iron and magnesium; it undergoes pressures from 10,000 to 1 million atmospheres and temperatures from 1,000° to 2,000°C. At a depth between 2,900 and 400 kilometers the mantle is very viscous, but above that it is relatively fluid, especially in the last 100 kilometers; this "fluidity" is expressed by giant convection cells similar to those that agitate boiling water: the liquid, heated in depth, rises, cools at the surface, becomes heavier, and sinks, and then the cycle begins again.

On this upper part of the fluid mantle floats, like a plank on water, the third envelope of the earth, the lithosphere; it is between 80 and 120 kilometers thick, and is solid, consisting of basalt closely resembling the mantle in mineralogical composition. It is subject to pressures from 1 to 10,000 atmospheres and to temperatures from 0° to 1,500°C. The earth's crust, lighter, granitic in nature, and rich in silicates of alumina, forms part of it; it is between 10 and 50 kilometers thick, and forms the five continents encased in the lithosphere like wooden planks frozen in ice.

The lithospheric envelope is divided into six principal plates like the joining elements of a gigantic puzzle. These plates slide over the fluid part of the mantle and are dragged along by the convection cells; they drift, knock against each other, separate, or overlap in the course of millions of years like the blocks of ice of an enormous ice floe. These six principal plates have been named America, Eurasia, Africa, India, Pacific, and Antarctica. Each is perfectly rigid; should Paris be displaced toward the east, Moscow and Peking undergo the same movement, since these three capitals find themselves on the same plate, are passengers on the same raft—the Eurasian plate. Only the joints, the boundaries between plates, are active, and they are the seat of separations, shocks, and overlappings; most earthquakes and volcanic eruptions are concentrated there (figure 3). They are zones of weakness, the scars of the lithosphere, like the cracks in a piece of steel that yield to the slightest stress. Two of these boundaries between plates are famous: the Pacific ring of fire, which surrounds that ocean with garlands of active volcanoes, and the Alpine chain, rich in earthquakes, which winds from Western Europe to Indonesia by passing through the Himalayas. For nearly a century scientists have been asking themselves why earthquakes and volcanoes should be localized in the narrow lines winding at the surface of Earth. And now finally the mystery is solved: these lines are the boundaries between plates. But how do they appear on the land and how do they function? Here too modern geophysics supplies an answer, in this case through the study of the ocean floor so long neglected by researchers.

By plowing the seas while aboard oceanographic ships, geophysicists have noted that two particularly characteristic reliefs exist at the bottom of the ocean. On the one hand, there are chains of undersea mountains, the mid-ocean ridges, which stretch for thousands of kilometers. From 2,000 to 3,000 meters high, 1,000 kilometers wide, they are sunk in their middle, throughout their entire length, into embanked valleys 20 to 50 kilometers wide and 1,000 to 2,000 meters in depth. On the other hand, there are oceanic trenches, arc-shaped, several thousand meters deep, and hundreds of kilometers long.

The best known mid-ocean ridge is the one that rises in the middle of the Atlantic. The important oceanic trenches are disposed around the Pacific; they are bounded both by strings of islands—volcanic arcs such as the Japanese and Indonesian archipelagoes—and by colossal mountain chains such as the Andes. These two types of reliefs are only the illustration at the surface of the boundaries between plates which function according to a simple mechanism discovered and then explained in the last decade: the material that constitutes the lithospheric plates comes from the upper fluid mantle in the form of magma, which mounts toward the surface thanks to the rising movements of the convection cells. It clears a passage for itself in the large fractures situated at the bottom of the deep valleys of the mid-ocean ridges, gives rise to submarine volcanoes, expands the earth, cools, and becomes welded with the edges of the plates bordering both sides of the magmatic injection. Later a new intrusion of molten matter cuts the previous injection into two equal parts, separating them symmetrically to either side of the axis of the mid-ocean ridges at an average speed of 1 to 12 centimeters a year. This mechanism of sea-floor spreading proceeds during tens of millions of years; the plates are slowly pulled apart from the axes of the ridges, then reach the zones where the descending movements of the convection cells draw them into the depths of the oceanic trenches, where they sink beneath other plates, penetrate the mantle, and are entirely melted at 700 kilometers underground.

Thus, along certain lines, certain plate boundaries, as in the middle of the Atlantic, in the East Pacific, and the Indian Ocean, the earth expands, lithospheric matter is formed; along other boundaries, this surplus of matter is resorbed, melted, and the earth contracts, as in the Pacific ring of fire or the Alpine chain. The continents, caught in the lithospheric plates, are displaced like objects on rolling carpets; they are too light to be drawn into the trenches and remain on the surface, like scum, floating and folding themselves to produce mountain chains.

The theory of plate tectonics allows us to state, for example, that New York and London are becoming more separated by an average of 3 to 4 centimeters a year, since these two cities are situated on two different plates, the American and the Eurasian,

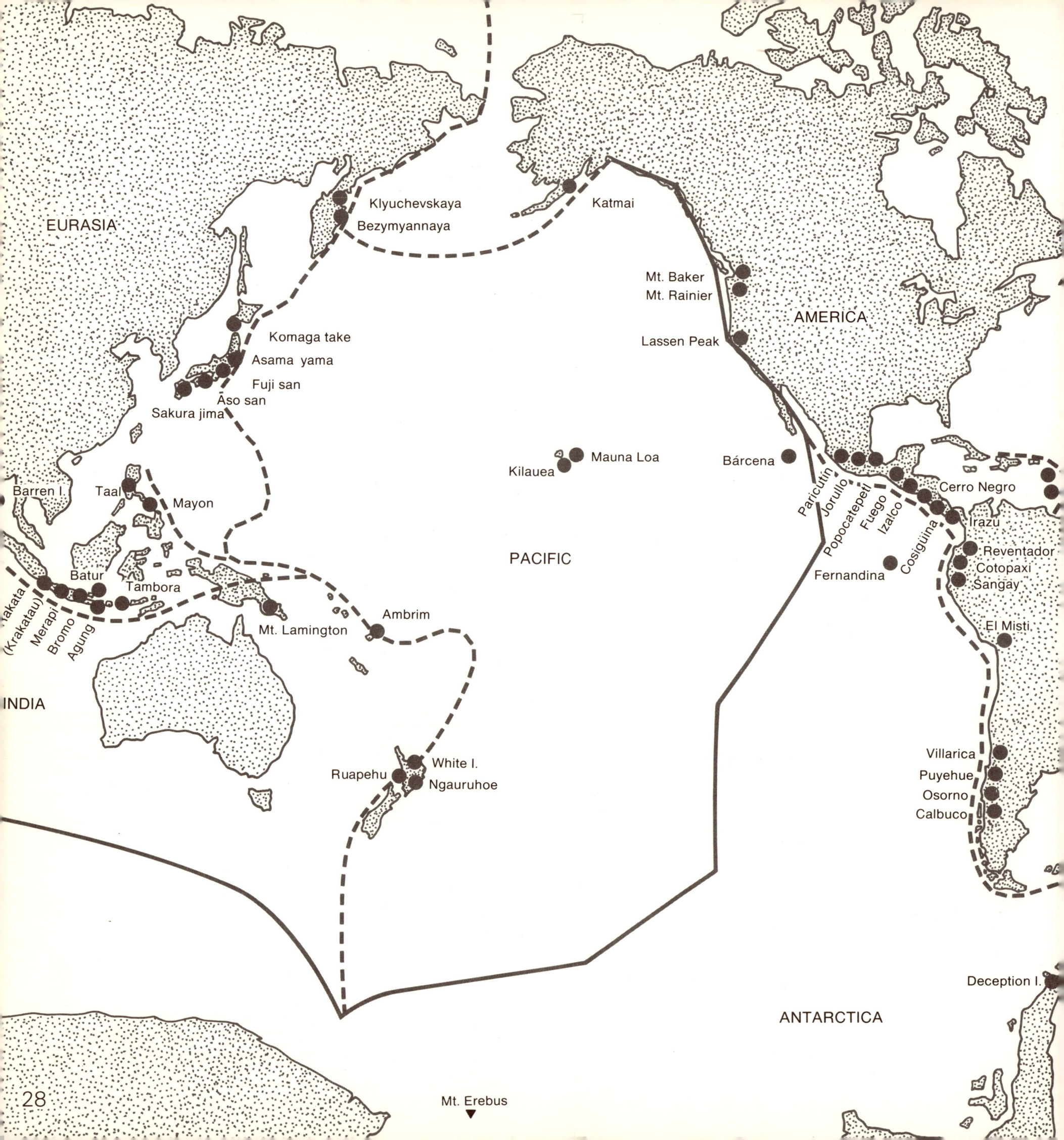

EURASIA

Klyuchevskaya
Bezymyannaya

Katmai

Komaga take
Asama yama
Fuji san
Āso san
Sakura jima

Mt. Baker
Mt. Rainier

AMERICA

Lassen Peak

Barren I.
Taal
Mayon

Mauna Loa
Kilauea

Bárcena

Paricutín
Jorullo
Popocatepetl
Fuego
Izalco
Cosiguina

Cerro Negro
Irazu

Reventador
Cotopaxi
Sangay

PACIFIC

Fernandina

(Krakatau)
Merapi
Bromo
Agung
Batur
Tambora

Mt. Lamington

Ambrim

El Misti

INDIA

Villarica
Puyehue
Osorno
Calbuco

Ruapehu
Ngauruhoe
White I.

ANTARCTICA

Deception I.

28

Mt. Erebus

Jan Mayen

Askja

Hekla
Surtsey Heimaey

EURASIA

Vesuvius
Faial Vulcano Stromboli Büyük Ağri Daği (Ararat)
Etna Thíra
(Santoríni)

La Palma
Pico de Teide

AFRICA

Mt. Pelée
Soufrière Fogo

Ertaale

Mt. Cameroun

Nyamuragira
Nyiragongo Kilimanjaro
Karthala

Piton de la Fournaise

Tristan da Cunha

Saint Paul I.

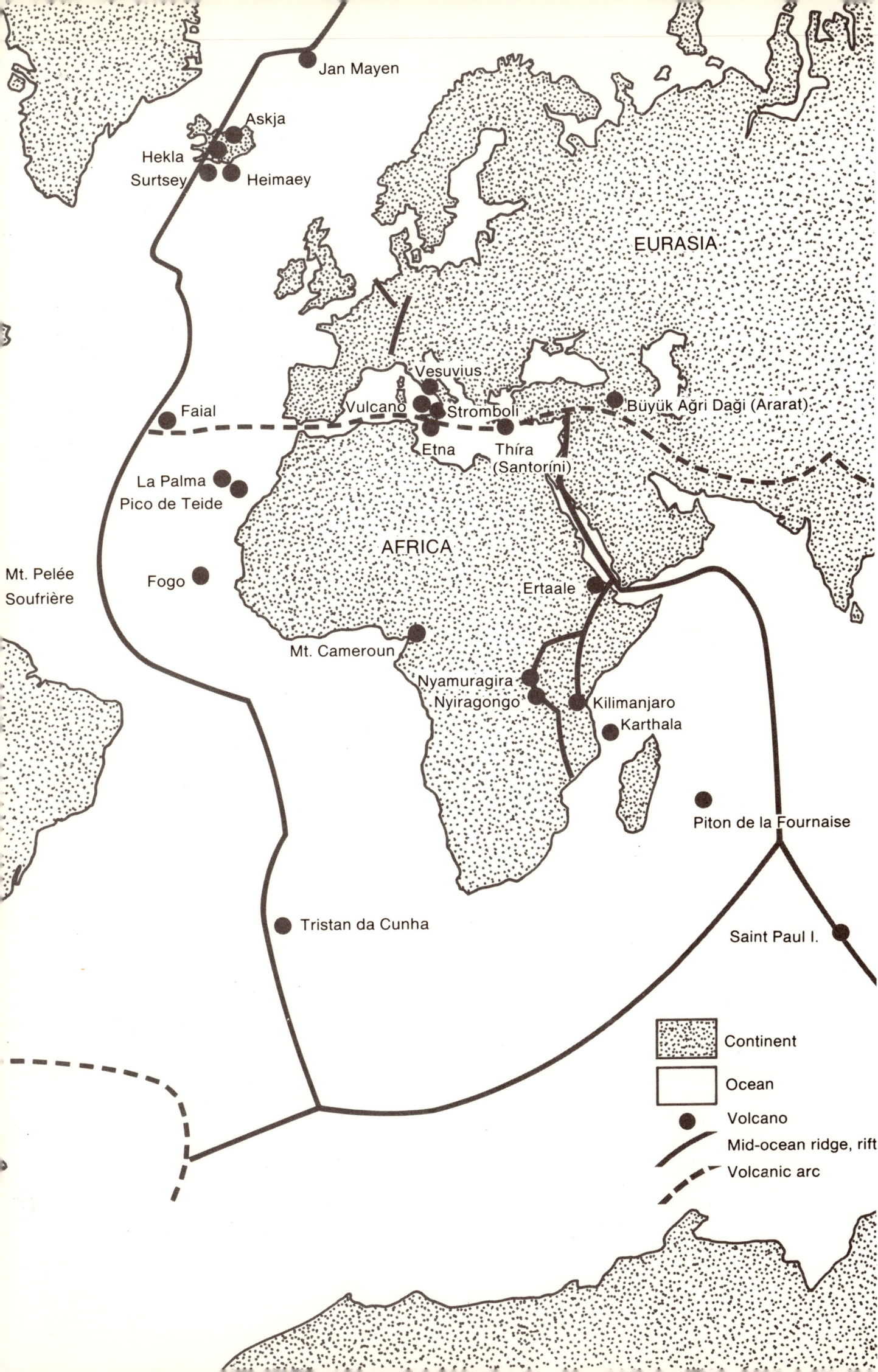

3
THE SIX PRINCIPAL
TECTONIC PLATES AND
THE PRINCIPAL VOLCANOES

separated by a mid-ocean ridge which pulls them apart. On the other hand, Algiers is moving closer to Marseilles by an average of 2 centimeters a year, the Mediterranean narrowing, while the African plate sinks under the Eurasian plate all along an oceanic trench that is quite visible south of Crete. Likewise, the Hawaiian Islands, situated in the middle of the Pacific, are being displaced toward the northwest by an average of 7 centimeters a year, and if this drift continues for some tens of millions of years, these islands will be engulfed in the trench of the Aleutians or that of Japan.

The geological forces of the continental drift are probably the convection cells of the upper mantle, due to the fact that the earth is a hot planet; 80 percent of its heat comes from nuclear reactions of radioactive elements such as uranium, thorium, and potassium contained in the rocks; the other 20 percent is calories preserved inside the globe since its creation. It took two generations of scientists to concretize and get this dynamic rather than static image of the earth accepted. The most important consequence of this revolution in the geological sciences can be immediately seen: volcanism plays an essential role in the evolution of the skin of our planet. At the level of mid-ocean ridges, numerous fissural volcanoes, essentially effusive, are aligned in the broad encased valleys that follow the axis of the submarine mountains (figure 4). On the edges of the oceanic trenches, a multitude of explosive volcanoes is strung along volcanic arcs. If it is true that the majority of the world's volcanoes is located on the ridges and arcs, certain volcanic structures are, however, very far away from them, such as those of East Africa, which are scattered through a gigantic collapsed valley, a rift, that winds from Mozambique to the Dead Sea through Tanzania, Zaire, Kenya, Ethiopia, and the Red Sea, and that moreover extends as far as Oslo through Greece, Italy, the Rhone Valley, and the plain of the Rhine. This immense crack in the heart of the continents is in the process of opening, of expanding to give birth to a new boundary between two plates that, later on, will perhaps become a mid-ocean ridge. At the bottom of the rifts arise numerous effusive and explosive volcanoes resulting from repeated injections of magma into the earth's crust.

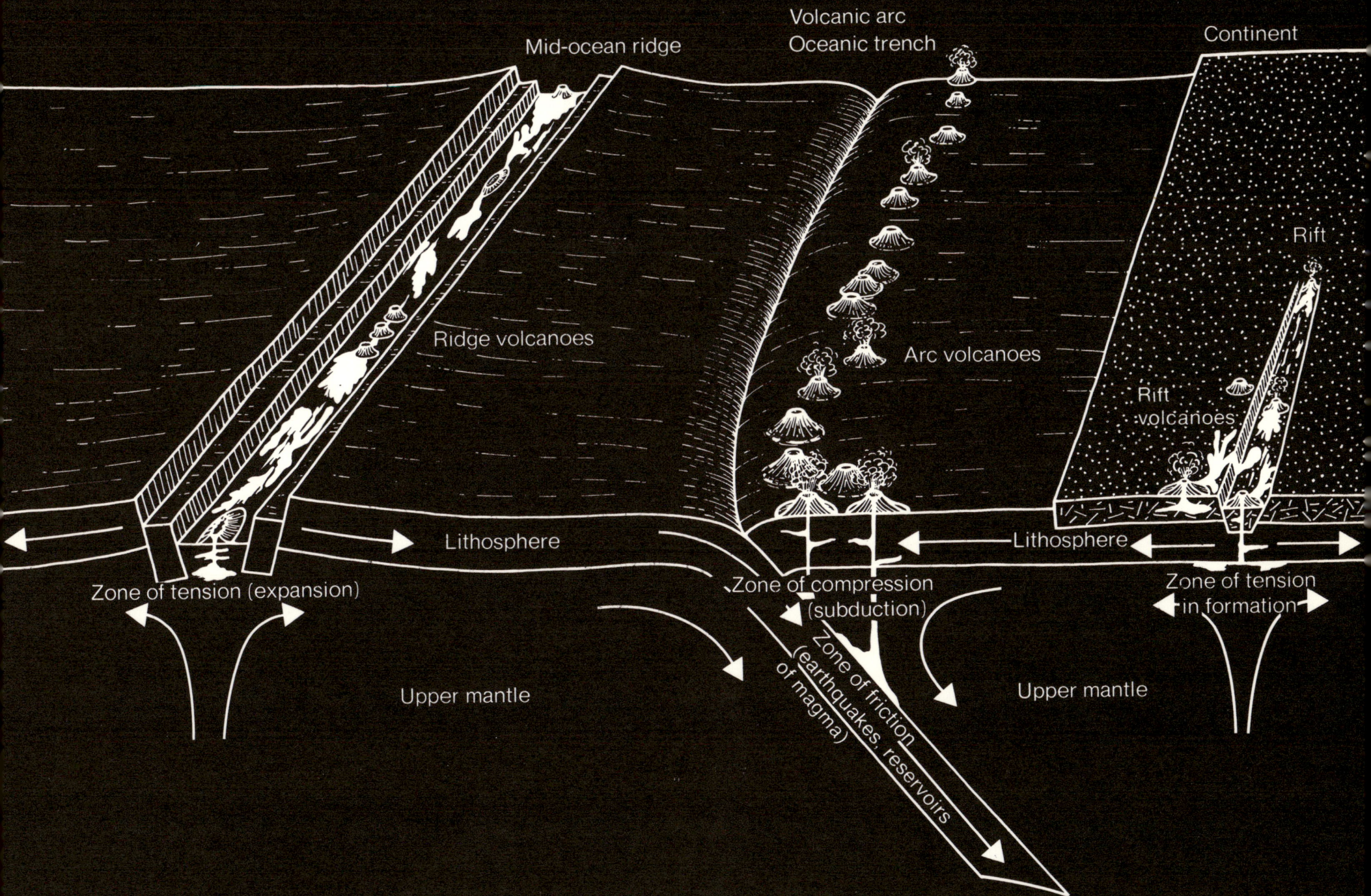

Mid-ocean ridge

Volcanic arc
Oceanic trench

Continent

Ridge volcanoes

Rift

Arc volcanoes

Rift
volcanoes

Lithosphere

Lithosphere

Zone of tension (expansion)

Zone of compression
(subduction)

Zone of tension
in formation

Upper mantle

Zone of friction
(earthquakes, reservoirs
of magma)

Upper mantle

4 THE THREE TYPES OF VOLCANISM IN THE LIGHT OF THE PLATE TECTONICS

(For simplification the correct proportions of thicknesses have not been followed in this diagram.)

31

The First Spasms of a Volcano in Activity
Engraving by Edouard-Yan Dargent

THE MECHANISM OF VOLCANOES

Rather than seeking a definition, necessarily imperfect, of a volcano, we will attempt to explain its mechanism, from the depths of the earth to the front of the lava flow that becomes immobilized on the flank of a volcanic mountain.

The upper mantle and the lithosphere are constantly subject to colossal stresses that provoke readjustments in the earth and fold and fracture the rocks, producing earthquakes, enormous frictions, and significant lowerings of pressure. Very great quantities of energy are thus liberated; this energy is converted into calories, which are added to those released by nuclear reactions due to the radioactivity of the earth, which are melting the rocks to give birth to magma. The hot points thus created take the form of reservoirs of melted silicates, containing crystals in suspension and volcanic gases in solution. Their temperature varies from 500° to 2,000°C, their pressure from a few hundred to more than 200,000 atmospheres. They are located between a few kilometers and 700 kilometers underground. For example, the reservoir of Vesuvius is 5 kilometers deep, that of Klyuchevskaya in Kamchatka is at 70 kilometers, and those of Kilauea in the Hawaiian Islands from 8 to 60 kilometers under the surface. If the reservoir is superficial, situated in the earth's crust, it is most often of acid nature, which means rich in silica, since it results from the fusion of granitic rocks, themselves very rich in this compound. On the other hand, when the hot point is situated under the earth's crust, in the lower part of the lithosphere or in the upper mantle, it is composed of basic magma, poor in silica because it results from the fusion of basalt or peridotite, rocks with little silica. This distinction between an acid and a basic magma is of great importance in volcanism. It determines, on the surface, important differences in the types of eruptions.

Once the reservoirs are formed, the magmas develop in the course of time and differentiate themselves. The heavier crystals descend to the bottom of the reservoir, the lighter ones form a scum that floats on the denser magma. Gases circulate in the melted matter, bringing volatile substances toward the upper part. Elsewhere, along the walls of the reservoir, the magma at high temperature melts the enclosing rocks, assimilates, and digests them. Little by little, the magmatic composition changes; it is no longer homogenous in the whole reservoir. These melted masses often have impressive volumes: the reservoir of Klyuchevskaya has 20,000 cubic kilometers, that of Vesuvius 50 cubic kilometers.

Then, thanks to a zone of weakness in the enclosing rocks, the magma starts its ascent toward the surface; it penetrates like a wedge into the fissures, opens and tears them, and little by little proceeds by fits and starts toward the place of eruption. This ascent—from 500 to 1,000 meters a day, according to certain volcanologists—is accompanied by crackings, by small earthquakes. When the hot magma under pressure arrives near the surface, the ground rises, expanding slightly while at the same time slowly warming up. Earthquakes, deformations of the ground, and the appearance of hot points on the surface sometimes allow volcanologists to predict the eruption. As early as 1942, scientists of the volcanological observatory on the island of Hawaii predicted an eruption of Mauna Loa several months in advance; according to them, it was to happen on the northeastern flank of the volcano at an altitude of between 3,300 and 3,600 meters; it began on April 28, 1942, on the foreseen spot, at 3,400 meters.

The progress of the magma in the earth is often complicated; two kinds of conduits predominate: chimneys or "necks," which are roughly cylindrical; and "dikes," long fissures filled with lava. After this long journey with all its obstacles, the magma reaches the surface, a fissure opens, and the sudden decompression of the gases dissolved in the molten matter triggers the eruption.

Three types of materials then spurt from the fissure: gases, lavas, and pyroclastic products.

The gases, causes of the eruption, escape from the magma in fumaroles. They have temperatures of between $100°$ and $1,200°C$, and are 90 percent steam, the rest being carbon dioxide, carbon monoxide, hydrogen, sulfurous gases, halogenous acids, ammonia, and traces of many other volatiles.

The origin of the steam is very controversial. According to some, it is juvenile water

Smooth lava (Pahoehoe)

Ropy lava (Pahoehoe)

Entrail lava (Pahoehoe)

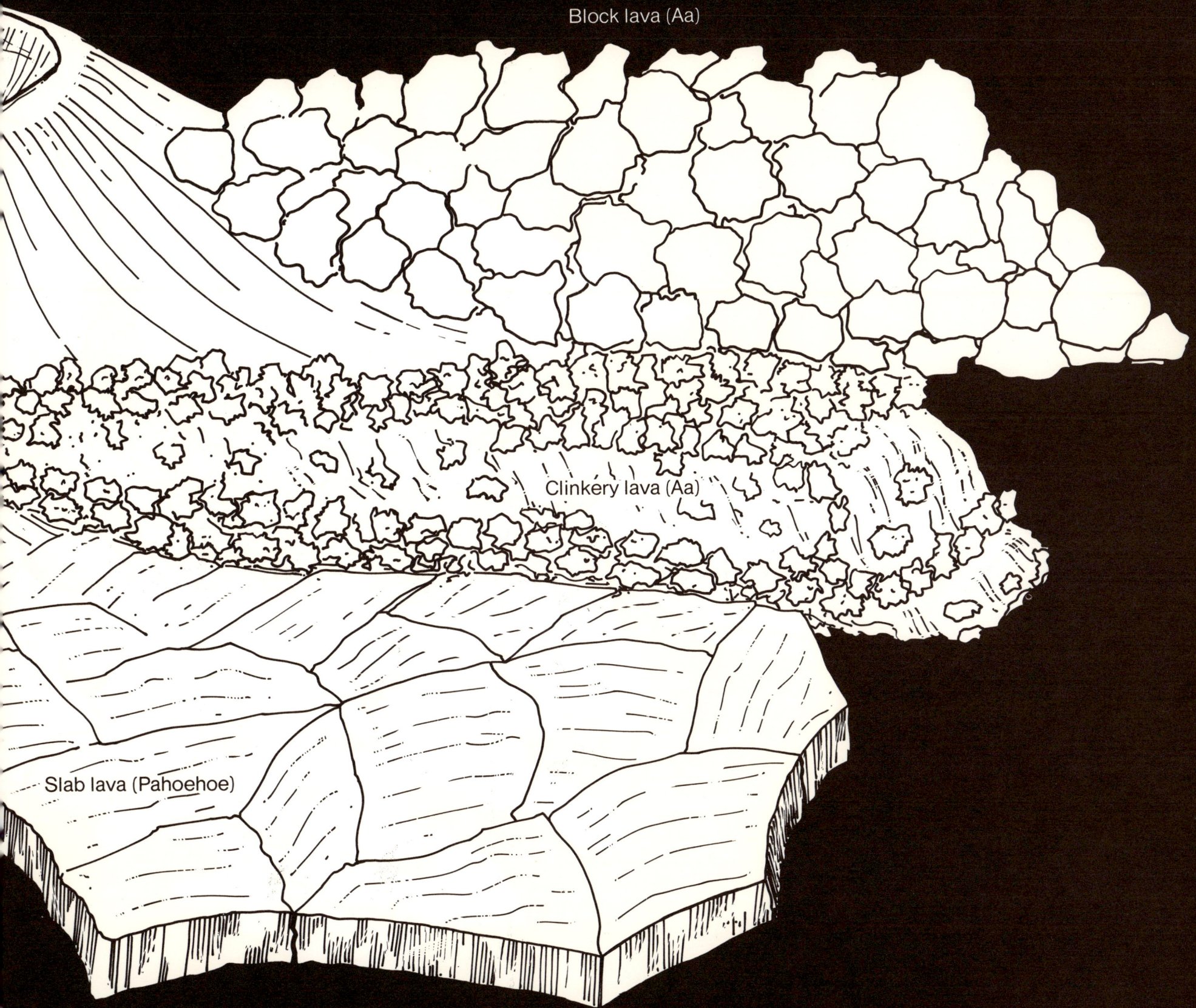

Block lava (Aa)

Clinkery lava (Aa)

Slab lava (Pahoehoe)

5   THE DIFFERENT TYPES OF LAVA FLOW

born from the fusion of the rocks; for others, it is only surface water infiltrated underground and heated in the depths. It is certain on the one hand that a cubic kilometer of granite, when melted, releases 7 billion cubic meters of steam; but on the other hand, volcanologists have observed that almost all volcanoes are near oceans and that seawater infiltrates the porous materials of volcanic structures. Many investigations would seem to show that both origins coexist.

The lavas, which are more or less degassed magmas, flow from the fissures at temperatures varying between 700° and 1,100°C; they contain silicates of alumina, iron, calcium, magnesium, sodium, potassium, and other elements in lesser quantities. Their flowing and cooling speeds at the surface depend on their chemical composition, temperature, gas content, richness in crystals, and on external conditions (the incline of the volcano, whether the flow is under the ocean or under the ice, and so on). When the lavas cool slowly, the crystals have time to form and to grow; but if they harden quickly, they do not crystallize but take the structure of glass.

Acid lavas, rich in silica, are most often very viscous; they accumulate around the exit orifice of the magma and build up domes; the volcanic rocks called rhyolite, trachyte, andesite, and phonolite are of this kind.

On the other hand, the basic lavas, poor in silica, are very fluid; they cover great surfaces and flow in superimposed streams, which, according to their appearance, have assumed evocative names: smooth, ropy, entrail, slab, clinkery, block, pillow lavas (figure 5). Their speed of flow varies from a few meters a day to more than 30 kilometers an hour. The basalts are the best known basic lavas, and the most common on earth. Later, when erosion has dismantled the lava flows, some magnificent columnar jointing will appear; this results from the cooling of the lavas when numerous retreat fissures are formed in the contracting matter.

Pyroclastic products, the last elements in the trilogy of volcanic materials, are expelled with force from the places of eruption, then fall around the mouths in the form of bombs, blocks, lapilli, and ashes that are more or less cooled. Their accumulation builds up cones with craters. They are composed of the same elements as the lavas but have been pulverized by the explosions. The very varied shapes of volcanic bombs have inspired volcanologists to give them significant names: spindle-shaped, cow-dung, breadcrust.

If gases, lavas, and pyroclastic products are simultaneously present in almost all volcanic eruptions, their importance and respective behavior vary to a considerable degree. Volcanologists have thus been led to create a classification of eruptive dynamisms based precisely on the nature of the materials emitted by volcanoes. Nine types of eruptions have been defined.

Hawaiian eruptions:

These are characterized by an abundance of very fluid basaltic lavas, which spout in fountains, spread in sheets, or bubble in a lake of lava. These lava flows give rise to enormous cones with very smooth slopes, endowed at their summits with large deep craters, and sometimes marked on their flanks by narrow collapsing structures: the rifts (figure 6).

Fissure eruptions:

As the preceding type, fluid basaltic lavas predominate; they flow out of long fissures, then spread over large surfaces. Explosive activity sometimes engenders dozens of small cones that align themselves on the cracks. This type of eruption is often accompanied by open fractures that do not emit lava and correspond to an aborted volcanism, the magma having expanded the earth but without having reached the surface (figure 7).

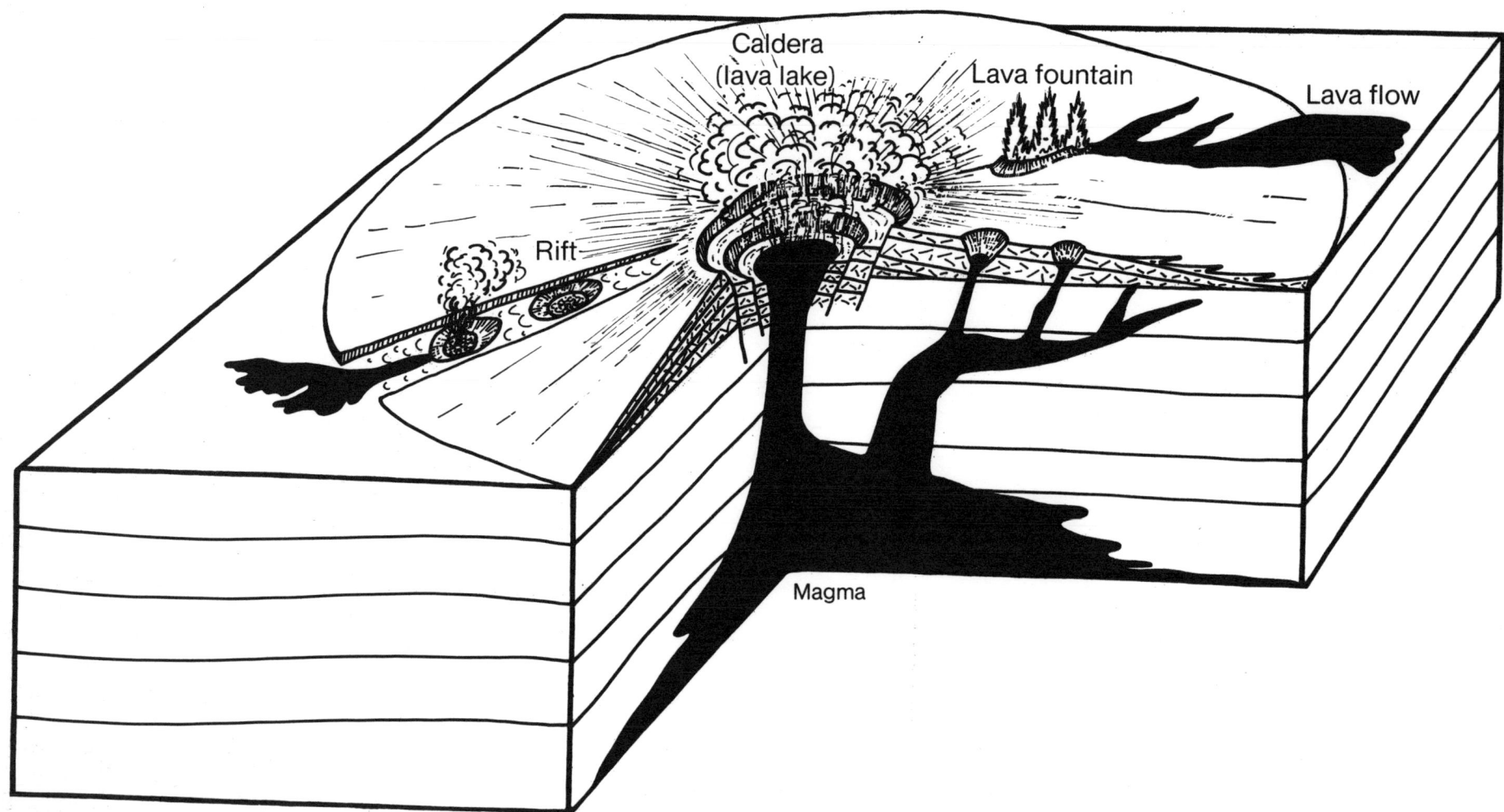

Caldera
(lava lake)

Lava fountain

Lava flow

Rift

Magma

6   HAWAIIAN ERUPTION

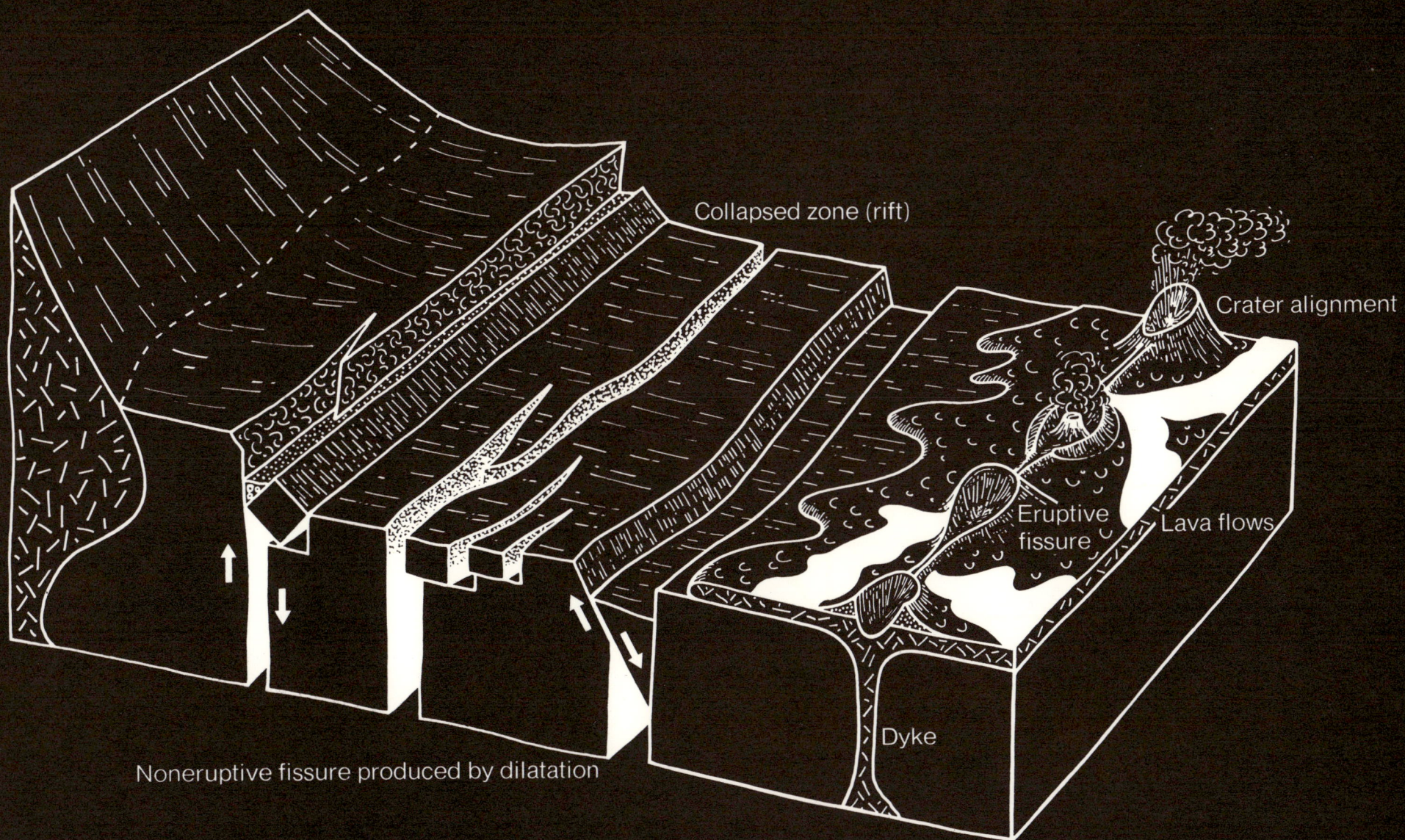

Collapsed zone (rift)

Crater alignment

Eruptive fissure

Lava flows

Dyke

Noneruptive fissure produced by dilatation

7   FISSURE ERUPTION

Strombolian eruptions:

Strombolian activity is manifested by the alternation of explosive phases with the ejection of molten products and effusive phases with the emission of streams of lava. A stratovolcano builds up little by little, the strata being alternately rich in pyroclastic products and in lava flows. The volcanic structures that result from this type of eruption are cones with simple craters, joining, one inside the other, or taking a horseshoe shape (figure 8).

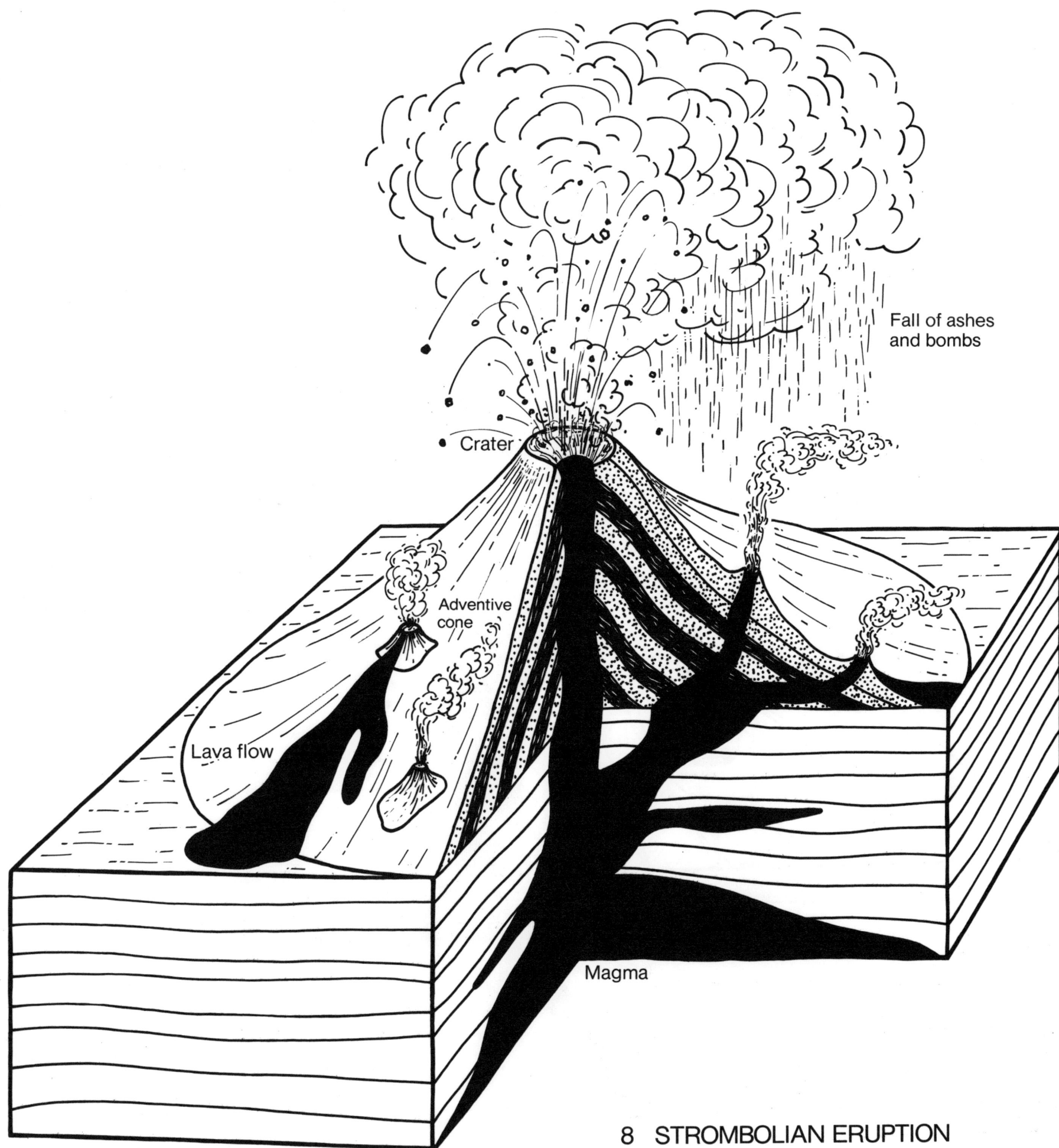

Fall of ashes and bombs

Crater

Adventive cone

Lava flow

Magma

8 STROMBOLIAN ERUPTION

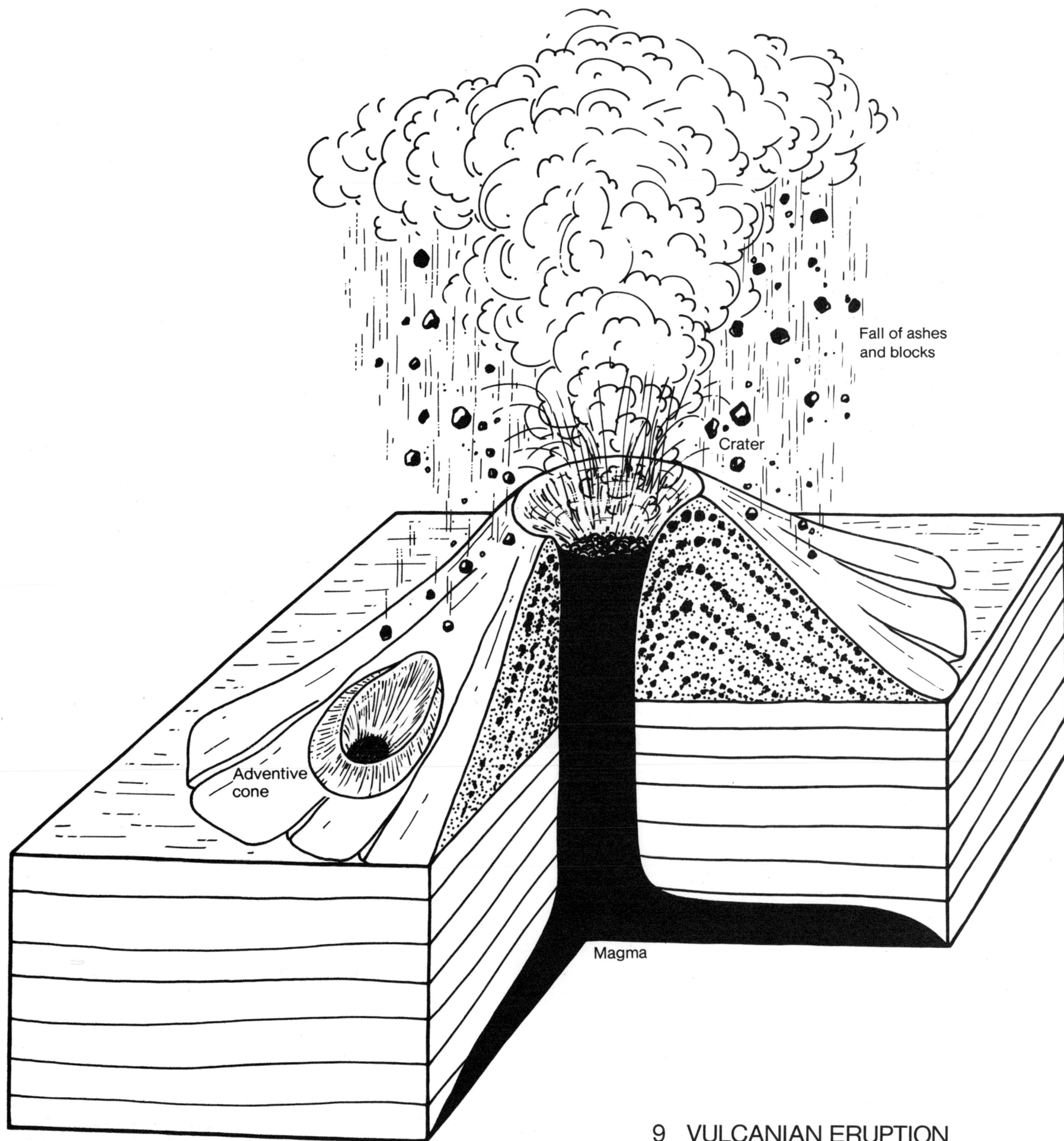

Fall of ashes
and blocks

Crater

Adventive
cone

Magma

9 VULCANIAN ERUPTION

Vulcanian eruptions:

When a magma is acid, it is viscous and opens a passage for itself through the earth only with great difficulty. Often it obstructs the chimney so that gases accumulate under the magmatic cap. The pressure grows to such an extent that a strong explosion occurs, expelling ashes, breadcrust bombs, and old blocks from the crater. During their ejection, these materials are almost solid. The significant shape of this type of eruption is a cone with a crater filled with ashes and with blocks of all sizes. Streams of lava are rare in this case, with the exception of small flows of obsidian, "the glass of volcanoes" (figure 9).

Ultravulcanian eruptions:

These are characterized by a great abundance of gas in the magma. Enormous pressures accumulate and are released in violent explosions. This type of activity only lasts for the time of an explosion, which abruptly cuts the earth and violently projects ashes, bombs, and blocks. A circular explosion crater surrounded by a crown of pyroclastic products is formed. Flows are nonexistent. It frequently happens that the products ejected at the time of the explosion do not contain the least trace of magma, only old crushed rock. These paroxysmal explosive eruptions engender very large craters: explosion calderas (figure 10).

Dome eruptions:

In certain volcanoes, the magma is very acid, excessively viscous: it cannot spread out at the outlet of the eruptive mouth, but rises above the chimney to build up a pelean dome or a spine which gradually crumbles during its slow ascent. Very thick and short flows are sometimes emitted at the foot of the dome (figure 11).

Fumaroles

Explosion crater (maar)

10 ULTRAVULCANIAN ERUPTION

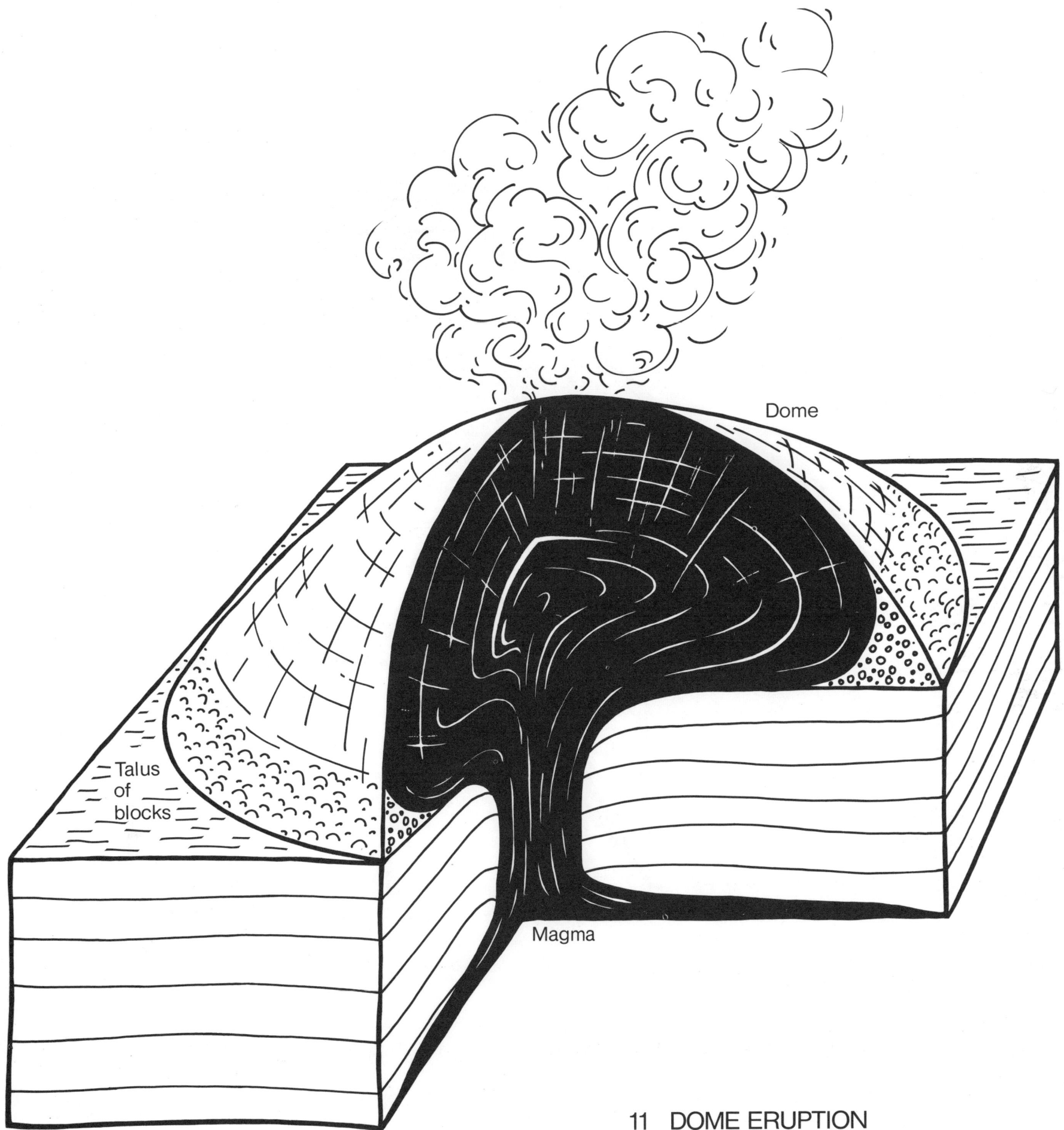

Talus
of
blocks

Dome

Magma

11 DOME ERUPTION

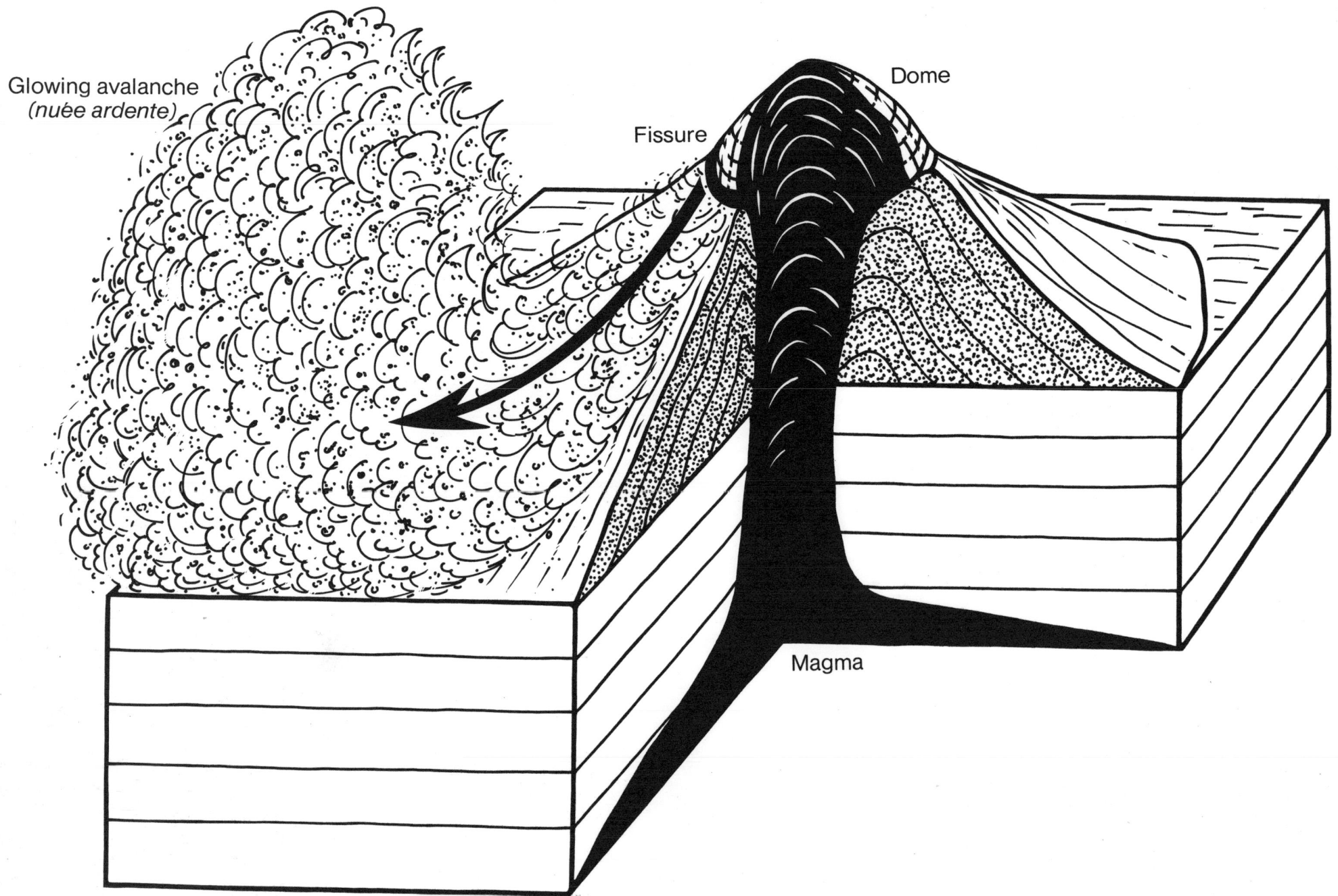

Glowing avalanche
(nuée ardente)

Fissure

Dome

Magma

12   PELEAN ERUPTION WITH GLOWING AVALANCHE

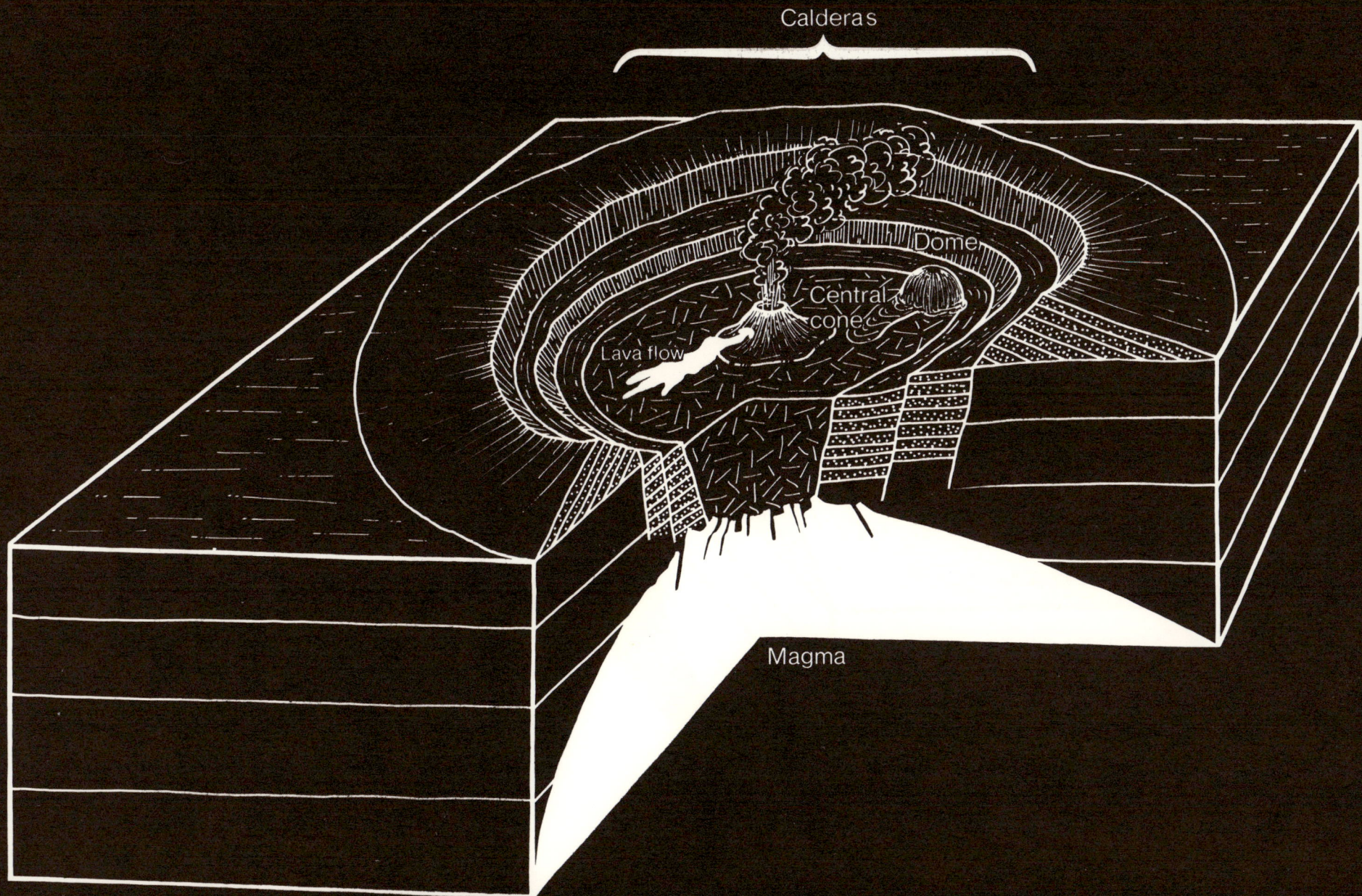

Calderas

Dome

Central cone

Lava flow

Magma

Pelean eruptions with glowing avalanches (nuées ardentes):

When the pressure of the gases under a pelean dome becomes enormous, a fissure opens abruptly at the base of the viscous lava cap: a glowing avalanche emerges. It is a kind of aerosol spray of gases, of small drops of molten acid lava and blocks of all sizes, which descends the slopes of the volcano at more than 100 kilometers an hour and at temperatures of several hundred degrees, destroying everything in its path. The burning clouds spread over large surfaces, level the relief, and produce deposits several meters thick, rich in small flattened drops of lava and in debris of welded crystal: the ignimbrites (figure 12).

Collapse eruptions:

Through successive eruptions, the reservoir of a volcano empties little by little, the roof of volcanic rocks covering it then collapses to fill the void, and a large depression appears at the surface of the ground: this is a collapse caldera. Many calderas are often one inside the other, resulting from successive collapses. A final stage of this type of eruption is characterized by the emission of a little quantity of magma at the bottom of the caldera with the formation of small volcanic edifices (figure 13).

Submarine eruptions:

When a volcanic eruption is produced undersea at a depth of more than 2,000 meters, no manifestation is noticed at the surface of the ocean, since the pressure of the water on the magma is too strong and prevents all explosive activity. On the other hand, when a volcano is near the surface of the ocean, the contact of the lava at more than 1,000°C with the cold seawater entails very violent exothermic reactions, which provoke explosions of cypressoid jets in the form of rooster tails. The molten rocks burst, break in pieces, are pulverized; structures develop in the flows in the form of bolsters: the pillow lavas. Often an emerged horseshoe-shaped crater is constructed, closing up later to isolate the magma from the seawater; the volcano's development then proceeds normally as on dry land. Volcanic islets are ephemeral; erosion and the vertical movements of the ocean floor make them disappear almost as rapidly as they have appeared. Their formation is often accompanied by tidal waves (figure 14).

A special type of eruption, that of volcanoes imprisoned under ice, involves the partial thawing of glaciers, thus launching devastating mudflows.

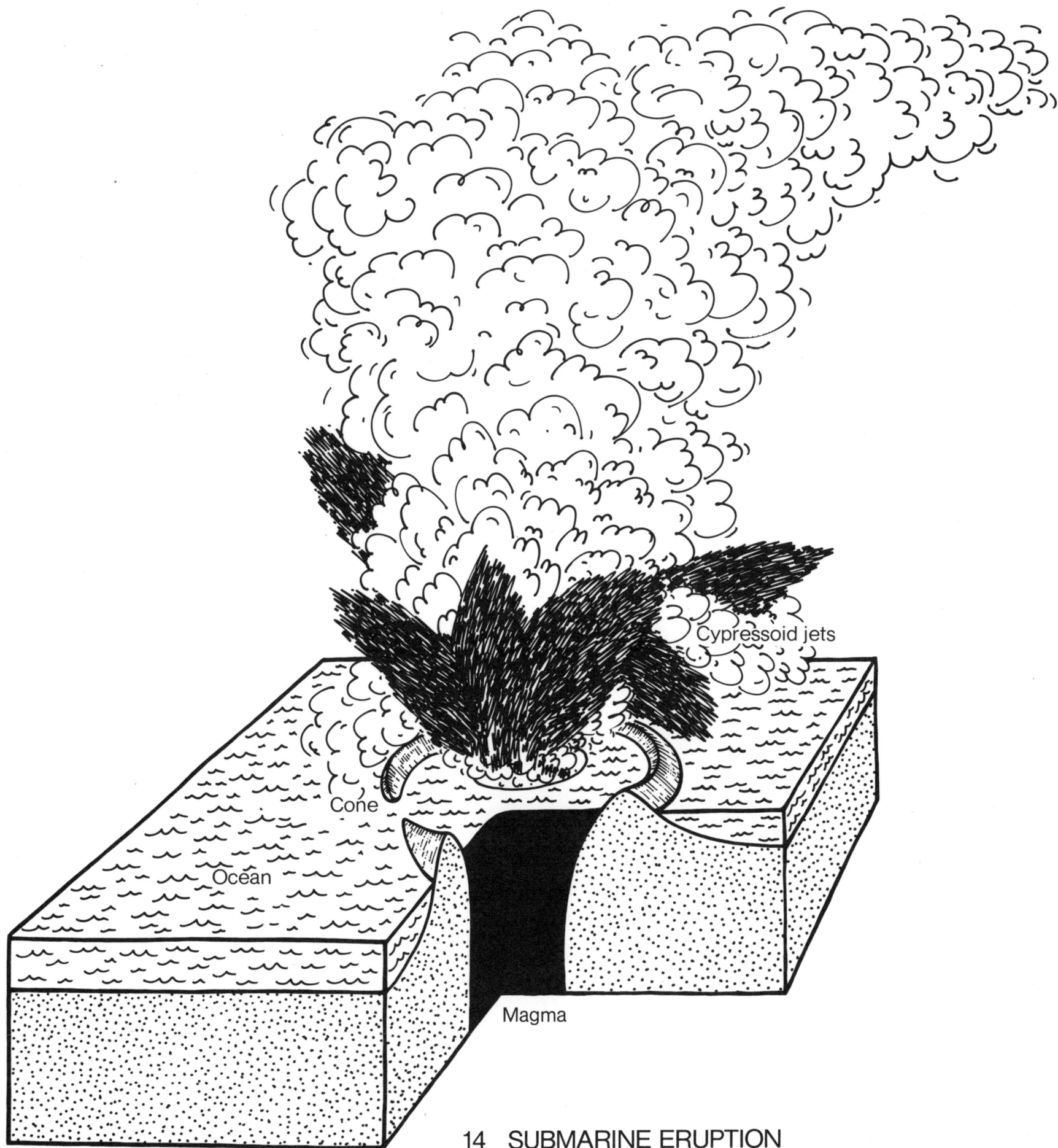

Cypressoid jets

Cone

Ocean

Magma

14 SUBMARINE ERUPTION

Volcanic eruptions are periods of fever in the life of a volcano; they are frequent, to be sure, but a volcanic mountain is in general more often asleep than active. In the course of these long sleeps, the secondary phenomena of volcanism are manifested: fumaroles, solfataras, hot springs, and geysers. They are the proof of the persistence of an underground hot point.

The genius of man now allows him to glimpse ways of utilizing the enormous quantities of calories imprisoned underground, by boring into the thermal zones to reach the hot flows of steam and water. Already the trains of northern Italy function by electricity produced from the geothermic energy of Larderello; already 70 percent of the Icelanders heat their homes by natural hot water.

All over the world, even in nonvolcanic regions, engineers seek to exploit the inexhaustible heat that is stored some kilometers under our feet.

On the other hand, we are still very far from being able to collect the enormous energy of a volcanic eruption. That of Krakatau, in the Sunda Strait, released in a single explosion on August 27, 1883, the equivalent of twenty thousand Hiroshima atomic bombs.

Hope exists, however, since the volcanologist is able to predict certain eruptions; he has even managed to arrest a lava flow on Heimaey in Iceland. So one day he may be able to master volcanic energy.

A serious problem remains: no rigorous criterion allows us to say that a volcano is definitely extinct, and at any moment a sudden raging of the planet Earth can launch the grandiose, mysterious, and exaltant spectacle that is a volcanic eruption.

MAX GERARD

The Possessed Mountain

Because man, hanging on the thin basalt coat of a sphere whose nucleus is energy,
does not like to meet anything so irrational as he,
volcanoes have fascinated him from the beginning,
and he has willingly used his fragility
to unravel the mineral susceptibilities of this splendid evil.
Once gestures of magic were entwined around the crater,
today when they are becoming scientific the fascination remains.
Nowhere else
does reason so collide with the living image of the night of time.
What the possessed mountain spits
is the unconsummated, the virgin, the accomplice of origins,
the substance that already was when everything began.
How to observe objectively this initial crucible of creation?

◄ A Volcano Furiously Spitting Death
Engraving by Edouard Riou

Among other mountains, the volcano even when extinguished attends to its spells,
it remains the one that destroyed the previously knotted and regrouped the incoherent,
the monstrous torch that carried the substantial fire,
the one that was able to modify even the very weave of the created,
to mark matter to the depths of its heart and forever.
In places where the earth built with such insolence and immodesty,
where it devotes itself to the whims of its implacable logic,
man can attach himself, measure, probe, add and subtract,
while avoiding looking the monster in the face.
Even extinguished, it remains the hypnotic, the sorcerer,
as primitive religions know,
where only initiates can approach the lips of the god.
We know neither the day nor the hour
when it may suddenly erupt in the midst of the life tormenting its flanks,
ambassador of the depths,
the only bond between the heart of the globe and its surface,
it reveals the power of its immoderation
by mobilizing for its own use forces prodigious enough
to change the intensity of gravity around the eruption.

The will of the crater consumes even the laws of nature.
Never does the fire nor its sister the heat
reach such grandeur
as in their frolic with the flames of stone.
To the fires of the eruption someone then comes to burn his reason to ashes.
He climbs the burning flanks,
fascinated in his approach to vital forces,
probing a will of action more ancient than all needs.
Far from giving up, he wants to overcome his own terror,
to forget the ancestral fear that drove him to hide or prostrate himself.
For the feast of danger he has put on a suit of light,
moving ever closer to the orgasm that impregnates the earth,
searching unconsciously for the images of a primordial womb.
To love is to set fire to the object of one's desire, and the fire,
sole communion between matter and life,
rekindles the cycle of the Phoenix for the benefit of the volcano.

It burns so as to re-create,
the glow of fire becomes an embrace,
procreating clasp of magma and flame,
struggle of shining beauty
that destroys and rebuilds, tears and will mend, burns and will make green again.
Man attempts to measure this art with his instruments,
trying to imagine how far the cynicism of the creator can go
when he sacrifices the present to his creation.
The possessed mountain attracts us so much
because it is in our image, incandescence and smoke,
all that is needed to love the sleep of an extinguished volcano.

Sleep

Adriane, exposed by erosion, is a chimney of trachyte that appeared six million years ago, at the end of the Tertiary period (Ahaggar, Algeria).

It is an extinguished volcano,
great beast crouching
in the hollow of the days;
it accepts nonchalantly
the admiration of men.

The silhouette of the stratovolcano Sundoro rises to an altitude of 2,550 meters, in the middle of Indonesian rice paddies whose fertility is the result of the rapid weathering, in the hot and humid tropical climate, of volcanic ashes rich in calcium and potassium (Java, Indonesia).

These fingers of stone
pointing to the sky
are no longer
furious chimneys
through which the earth
once vomited its substance.
Age has stripped them
of their useless rock,
abandoning
these hieratic old men
to the dreams of time,
geological sentries
who keep watch and wait.

Trachyte spines stand in the Tandjete wadi. They arose slowly, five million years ago, like heavy pistons driven by the pressures of volcanic gases (Ahaggar, Algeria).

The Massif du Bory is situated in the ▶ collapse caldera of Piton de la Fournaise, 9 kilometers in diameter; it is crowned by two small calderas: the Bory and the Dolomieu or Brûlant. Eruptions are hawaiian, fissural, or of the collapsing kind; they tear the flanks and bottom of the calderas, and a very fluid lava flows; small cones of scoria dot the whole volcano (Réunion Island, Indian Ocean).

The craters too
are waiting.
Empty nests,
delivered wombs,
they welcome the beauty
of these ravishing
inverted lakes.
One is of night
and the other of hope,
eternal duality
of the volcano,
but beware
of the sleeping water!

Sometimes impermeable layers settle
at the bottom of sleeping craters; rain
or spring water then accumulates in
these depressions. Keli Mutu consists
of three explosion craters: the first
contains clear water; the second is
rich in sulfurous fumaroles that give it
an emerald-green tint; the water of
the third is colored in dark red by iron
salts (Flores, Indonesia).

Eruption

On the flank of Bory, in the caldera of Piton de la Fournaise, a gaping fissure is adorned by congealed lava; here the earth has opened abruptly and fountains of liquid lava at more than 1,000°C have burst forth, inundating the volcanic earth (Réunion Island, Indian Ocean).

One day,
the earth breaks open,
in honor
of a gigantic egg that hatches.
One day,
the incombustible earth
itself is kindled.
The rock becomes vapor,
the mountain becomes cloud,
it eats the sun,
and the cloud of stone
carries to all this news,
that here the earth is biting
and erasing itself.

On January 23, 1973, a fissure 1.8 kilometers long opened some 250 meters from the town of Vestmannaeyjar in Iceland; a volcano was being born.

The cloud of steam and ashes rose 8 kilometers; a ▶ lava flow advanced into the ocean, transforming the seawater into great volutes of vapor. No one had expected the eruption, since the youngest volcano on the island of Heimaey, Helgafell, had had its last awakening five thousand years ago (Vestmannaeyjar, Heimaey, Iceland).

Uproar
is added to the noise.
The earth snorts,
spits, stretches itself
with all the violence of its mass.
The eruption lives
by its own rapacity,
in it everything is consumed,
it is the uniqueness
of an instant.

Between 1963 and 1967 three islands were formed to the southwest of Vestmannaeyjar, on the axis of the mid-Atlantic ridge. Submarine eruptions are violent, since the contact of the cold seawater with the molten rock provokes very strong explosions; the water is transformed into white clouds of steam, the lava is pulverized, disintegrates into ashes and blocks, and rises in black cypressoid jets (Surtsey, Iceland).

When the volcano
is impatient to conceive,
it hurries life,
and geological cycles
collide.
Monstrous parturition
from which nothing is missing,
neither the contractions of labor
nor the cries of birth
nor the spurting
of a flaming amniotic fluid.

Vulcanian eruptions are characterized by violent
explosive activity. At nearly all hours heavy
clouds of ashes, blocks, and gas develop abruptly
and rise from the crater crowning the andesitic
volcano Semeru, 3,676 meters high (Java, Indonesia).

In August 1971, Batur, one of the two sacred ▶
volcanoes of Bali, was active. The crater ejected
ashes and fragments of lava. This photograph, a
time exposure taken at night, shows the parabolic
trajectory of the volcanic bombs (Bali, Indonesia).

79

Eruption,
bursting from the depths,
the elemental,
original matter,
impatient in its prison,
tries to scale the sky.
Aspiration to mount,
nostalgia for élan;
on the altar of a pagan cult,
the volcano offers itself
a fiery feast
in honor of its own future.
Here is Wotan's brazier,
Vulcan's furnace,
the forge of the Cyclops,
Satan's pyre!
Here is the first panting,
the birth of matter,
here the gods are stoking
the superstition of men,
here the times are coming
of violence and damnation!

Strombolian eruptions are characterized by the
alternation of explosions in the crater of the
volcano and emission of lava flows—such is the
case of Stromboli. The vents explode irregularly,
ejecting volcanic bombs impelled by gas. Less
frequently the streams of lava descend the Sciara
del Fuoco and fall into the sea (Lipari Islands,
Italy).

It is duality,
it is antagonism,
it is the good,
the fire that lights and forgives,
it is the evil that the demons stoke.
It is sacred,
the appetite of the god
in its most sensitive form.
It is the symbol of violence
and sacrifice,
purification and punishment,
charm and barbarity.

By its explosions of ashes and bombs, and by its streams of lava overflowing onto the island and into the sea, Eldfell emitted 250 million cubic meters of volcanic products during six months of activity from January to July 1973 (Vestmann-aeyjar, Heimaey, Iceland).

Eruptions of the hawaiian type are often charac- ▶
terized by the presence of a very fluid lava that is discharged in flows, spurts in fountains, bursts in fragments of melted rock at a temperature of 1,000°C to 1,100°C, or accumulates in a lava lake (Nyiragongo; Virunga Chain, Zaire).

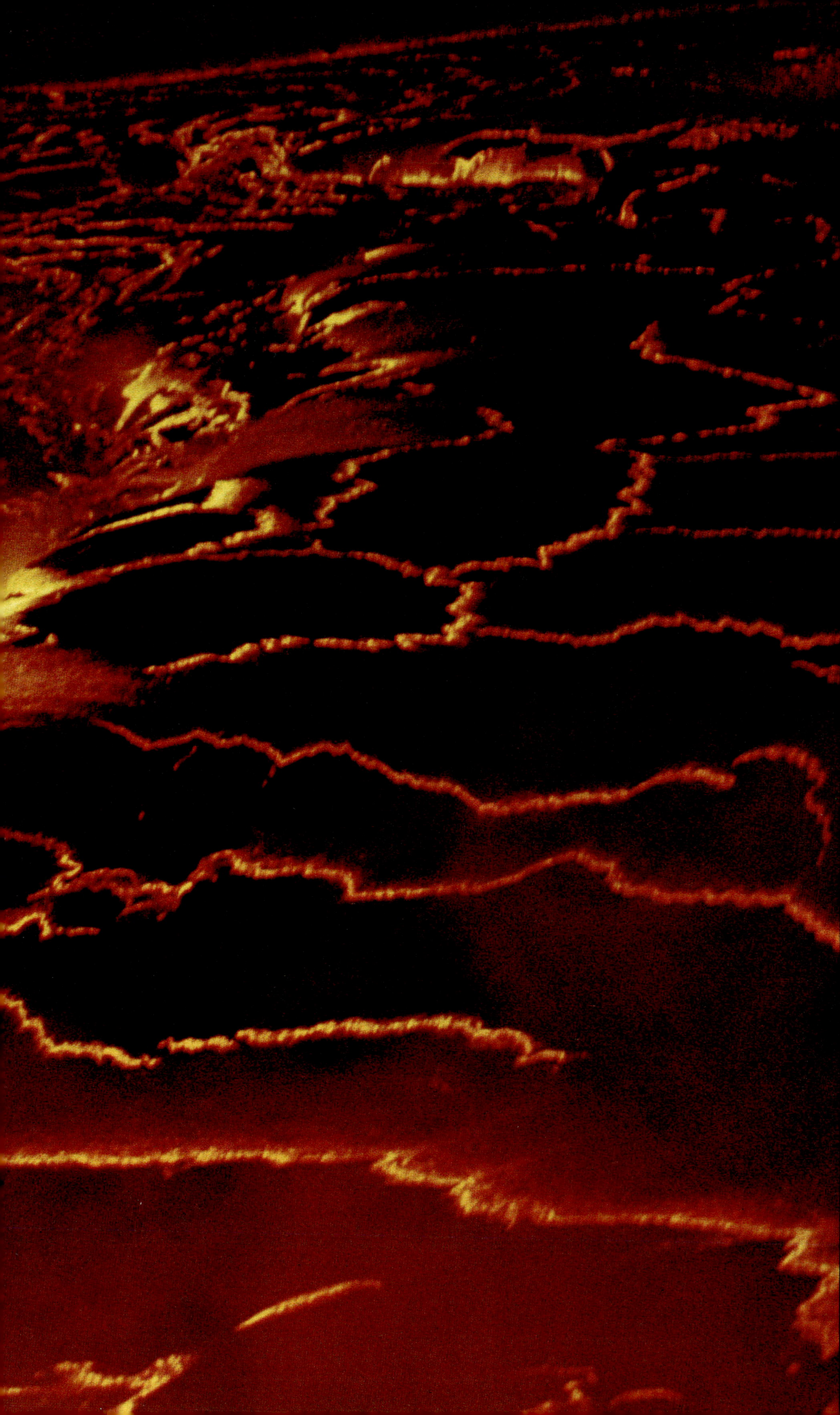

Reconsider the future
and reconsider space;
drunk with its power,
drenched with fire,
the lava advances.
Always the implacable
is allied with beauty,
and the horror of all times
reproduces the sacred,
when an earth spits its seed
while rediscovering
the deeds of its genesis.

◄◄ During the eruption of Rugarama, in April 1971, flows of liquid lava ran down the slopes of Nyamuragira, at a speed approaching 30 kilometers an hour. The speed of a flow is affected by the slope of the terrain, the temperature, the chemical composition, and the quantity of gas in the molten rock; it varies between a few meters a day to more than 30 kilometers an hour (Virunga Chain, Zaire).

The lava lake of Nyiragongo is 350 meters in diameter. It is covered by a film of hardened rock that becomes like a lithospheric plate over the molten rock at 1,100°C. The movements of the magma are often violent, and the lake sometimes resembles a sea of fire. In January 1972 the lake, which had just overflowed, ebbed and was re-engulfed in the central pit, where it was stirred by convection cells (Virunga Chain, Zaire).

93

Genesis of rites,
matriarchy of the goddess earth who uses fire
like the praying mantis uses her male
and copulates while devouring him,
here the jaws of the beast lie open.
What animal of the abyss,
what outcast monster
could have congealed its yawning in this stretch of lava?
While its skin becomes mineral,
while its fusion becomes stone,
with an all-new rock
nature repeats its lesson for us once more,
the only lesson it agreed to transmit to men:

> "Learn that everything begins again,
> that everything continues to begin again,
> that nothing has happened that is content to be,
> that the invariable could not prolong itself,
> and that from today's chaos
> a rock will emerge
> to catch the wind
> where a vein of earth
> will be born
> on which spring will be able to put its finger."

In the caldera of Piton de la Fournaise, at the time of its activity in 1972, streams of ropy lava formed at the base of the new cone, as a result of the flow of a thin and liquid lava, poor in gases, which cools quickly at the surface (Réunion Island, Indian Ocean).

This volcanic bomb, ejected by Eldfell in April 1973, is of basalt; it has a temperature of ▶ 1,000°C and comes from a depth of several kilometers (Vestmannaeyjar, Heimaey, Iceland).

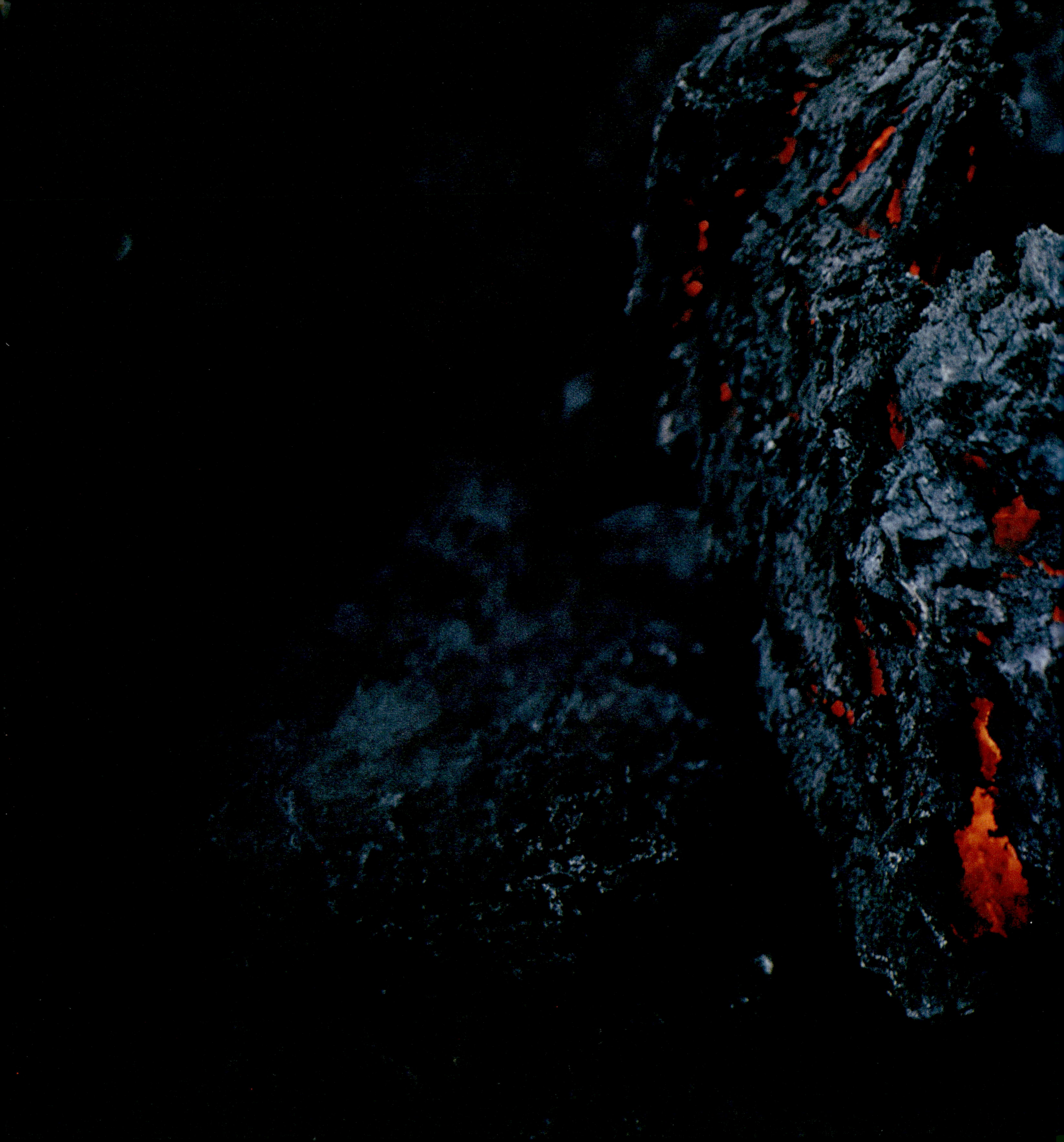

The bird of fire and cloud,
perched on the mountain,
has left a burning star
beside our magic spells.

From the top of Nyiragongo, 3,470 meters high, one can see the calderas of the volcano, the results of successive collapses. The lake of lava releases a great volume of steam and volcanic gases. In the distance rise two older eroded volcanoes: Karisimbi and Mikeno (Virunga Chain, Zaire).

After the fire, the damp,
after the ashes, the water.
It is the rearguard,
the sentry
placed by the volcano
to signal its conduct.
Harsh game of the heart
and of the shell,
seminal liquor,
the earth pursues
the attitudes of childbirth
so as to remind us
that nothing is consummated,
that the power remains,
which plots and waits.

Strokkur, at Geysir, functions by explosive
ebullition. At a certain depth in the pit of
the geyser, the water is in an overheated
state; that is to say, its temperature is
maintained above the boiling point thanks
to the pressure exerted by the column of
water. It takes only nuclei like bubbles of
gas or impurities for the liquid water to be
abruptly and massively transformed into
steam; a large steam bubble rises rapidly in
the pit and violently expels all the water of
the geyser. In 1974, this process occurred
about every five or six minutes at Strokkur,
the mixture of water and steam spurting to
a height of some 30 meters (Iceland). ▶

Fanatical cycle of the geyser that measures the feverishness of the globe.

The ancients,
who knew why
the earth
retains all the forces
of nature,
worshipped the vapors
that the soil improvises
and the mud breathing
through a thousand mouths.
Strange enchantment
exercised by the shores
where the surf
of the interior ocean
is coming to die.

Vulcano is sleeping since 1888, but gases escape in fumaroles, both in the crater and at the bottom of the sea, whose surface is sometimes covered with myriads of small bubbles as though to remind man that he is over a volcano given to sudden and violent eruptions (Lipari Islands, Italy).

When
at the surface of its skin
the earth
is feverish,
someone
should tremble!

In the thermal zones of Minahassa, there are numerous pools of boiling mud; they have average temperatures of 90°C. The mud, the result of a fumarolian weathering of old volcanic rocks, is heated by gases and steam coming from the depths. This is the solfatarian stage of an old volcanic zone (Celebes, Indonesia).

Someone

To protect himself from heat radiation rising above the lava lake of Nyiragongo, a volcanologist of the Vulcain Team, who is observing the continual stirring movements of the molten rock, wears fireproof clothing, made of asbestos covered with polished aluminum that reflects 90 percent of the heat radiation (Virunga Chain, Zaire).

Here solitude does not exist,
the volcano binds the man in intimacy.
It is the magical companion,
the fearful accomplice
in whose honor someone has put on
his best attire.
Anonymous seeker of the absolute,
in order to combat his fear,
he cocks an ear
toward the expressive earth,
feels the monster's pulse;
by pure conquest of his mind,
he clings to his knowledge
and reassures himself by studying.

At present it is possible to predict eruptions for many volcanoes, provided they are regularly surveyed and studied. In Java, the Vulcain Team sounds Merapi, the most regularly deadly volcano in Indonesia. Its pelean eruptions are characterized by the slow erection of domes of viscous lava from which glowing avalanches (nuées ardentes) sometimes escape. It was an eruption of this type that destroyed the town of Saint-Pierre in Martinique on May 8, 1902, Mount Pelée killing 28,000 people in a few seconds.

Two volcanologists of the Vulcain Team wear antishock fiber-glass helmets and fireproof clothing in order to approach the vents of Stromboli and install a gas analyzer under the rain of volcanic bombs (Lipari Islands, Italy).

Do the magic spell,
the beautiful roar
seek
better to deceive,
so as to incite man
to press closer,
to catch fire,
forgive,
and again
as always
make the gestures
of love?

A dome of viscous lava covered by a mass of dark blocks of hardened rock rises very slowly in the crater of Api Siau. Gases escape at a temperature of 1,000°C (Celebes, Indonesia).

But having been unable
to possess it,
someone planted his sword
in the skin of the dragon.
Derisive Saint George,
he yet knows
that his victory
will be called:
to Know!

A thermocouple continually measures
the temperature of this fumarole, at
180°C; the numerous substances
within the volcanic gases crystallize
gradually as the temperature drops;
white aureoles of boric acid and
ammonium chloride, bright orange
trails of iron chloride, and thin yellow
needles of sulfur are deposited
around the burning vents of Merapi.
Sumbing (3,371 meters), Sundoro
(2,550 meters), and Dieng (2,565
meters) pierce the sea of clouds with
their andesitic outline (Java, Indonesia).

Toward what strange shore
have they now sailed?
Navigators of vapors,
they obey the call
of the burning womb
that may shelter
the initial embryo.
Memory of a universe
damp and warm,
intrauterine nostalgia,
are they going to meet
the prodigious egg?
When one measures
a lake of impatience,
the only courage is
to know one's own fragility.

Kawah Idjen contains 36 million tons of sulfuric and hydrochloric acid, a million tons of potassium, 300,000 tons of alum, and 200,000 tons of aluminum. Members of the Vulcain Team navigate in an inflated boat on the lake of fuming acid in order to study it. This volcano sometimes awakens— the lake heats up, the acid overflows; the contents of the lake run down the slopes of the mountain carrying sand, rocks, and vegetation in giant flows of corrosive and destructive mud (Java, Indonesia).

Abundant sulfurous gases rise in the crater ▶ of Kawah Idjen and form deposits. The Indonesians make use of the sulfur (Java, Indonesia).

117

The volcano,
which has staged
its spectacle
at the very threshold
of men, then hurls
the great anathema:
    "Dust . . .
    I am the god.
    If I so desire,
    all will be nothing
      but ashes and dust."
Then only hope
or prayer remain.

◄◄ The eruptive fissure that opened on the island of Heimaey on January 23, 1973, was only 250 meters from the nearest houses of the town of Vestmannaeyjar, with 5,300 inhabitants. Volcanic bombs fell on the houses and set them on fire (Vestmann-aeyjar, Heimaey, Iceland).

Volcanic ashes rapidly covered the town of Vestmannaeyjar, transforming it into a northern Pompeii; but the Icelanders decided to combat the destructive force of the volcano, and, for the first time in history, men were able to arrest and divert a flow of lava, by pouring day and night 4,500 tons of seawater an hour on the molten rock at 1,000°C, thus cooling and congealing it (Vestmannaeyjar, Heimaey, Iceland).

At Batur one day,
some priests no longer had anything but their hands, hope, and prayer
to dam the sly appetite of their volcano.
Then they assembled
and knelt to offer their immobility to the sky.
They had decided that no monster would make them retreat any farther.
The lava advanced to where they knelt,
accepted their sacrifice,
and began to abduct them into the fire of its devouring passion.
But these monks had forged among themselves
the most formidable of weapons.
They were marching at the head of their faith,
the trust of a whole population threatened by the daughter of hell.
The burning mud struck that wall of hope,
never did it succeed in overturning it,
never again did it dare to invade the valley.
This monument should remind us all of the power of belief
and the grandeur of loving.
Thanks to love, the flame always engenders the seed,
fire of Prometheus animating the clay,
light of the Pentecost consecrating souls,
or pyre on which a prodigious bird is consumed,
to renew the cycle of nature, the cycle of the Phoenix.

In the opening of the temple door of Kintamani, built on the edge of the large caldera of
the Batur volcano, appears the *meru* with eleven piled-up roofs, covered with thatch;
this altar is dedicated to the two sacred volcanoes of the island, spirits of well-being
(Bali, Indonesia).

The Cycle of the Phoenix

Die and become,
maxim of the Phoenix,
the beautiful bird
that burns and is reborn
from its ashes
more dazzling
and more alive.
Lavish nature,
extravagant creation,
it effaces to rechisel
and does not even recoil
before the holocaust
of a volcano.

◀◀ Etna in Sicily is regularly active. In February 1974, at an altitude of 1,680 meters, the earth opened up, the lava gushed forth, a small cone of scoria was formed. Clinkery and block lavas advanced into the snow-covered pine barrens, burning thousands of trees. The lava flows are rapidly covered by a hard crust, but it often takes several months, even several years, for the flows to become completely cooled. A volcanologist has calculated that between the years 1500 and 1914 volcanoes have emitted 65 cubic kilometers of lava and 320 cubic kilometers of ashes throughout the world. The activity of a single volcano may last some hours, days, months, or thousands of years. Each explosion in Etna's crater releases energy equivalent to 500 tons of dynamite. After vegetation has been engulfed by ashes, some twenty years must pass before plants reappear on the ash cone (Italy). ▶

The burning earth
takes from the wise earth
its trees and its bushes,
its mosses and birds,
to make of each a flame.
Divergent image,
it requires
all the passion
of hope to be convinced
that the eruption
goes on repeating:
Die and become!

Here seek no longer
the consolation of a reason
nor an exercise in logic.
The fiery appetite
has silenced
all that was not
fury, irritation,
revolt, and unreason.
The volcano has closed again,
after having painted everything
gray, lifeless, black,
and somber blue.
Expressionist image,
in its gravity
the reign of stone begins.

In the old Kawah Upas crater of Tangkuban Perahu, carbon dioxide, heavier than air, is emitted. The toxic gas overtakes men and animals, who die by asphyxiation in these valleys of death (Java, Indonesia).

132

Over the rigor of the stone,
over its necessity
remains only silence,
the finest fruit
of patience.

The volcano-tectonic depression of Lake Assal, a lake with waters supersaturated with salt and gypsum, is a recent boundary of two lithospheric plates. The blocks of basalt are split and rounded by the heat and wind of the desert climate (French Territory of Afars and Issas).

Lava congeals in multiple forms. When it is very ▶ fluid, as at Nyiragongo in Zaire, it spreads out over large surfaces in thin sheets under which the lava forms delicate stalactites; sometimes the molten rock spurts in incandescent waves against the walls of the crater, and these "sprays" fall in pearls and drops of lava. When still-fluid volcanic bombs revolve in the air, they take the form of helixes, as at Askja in Iceland during the eruption of 1961.

135

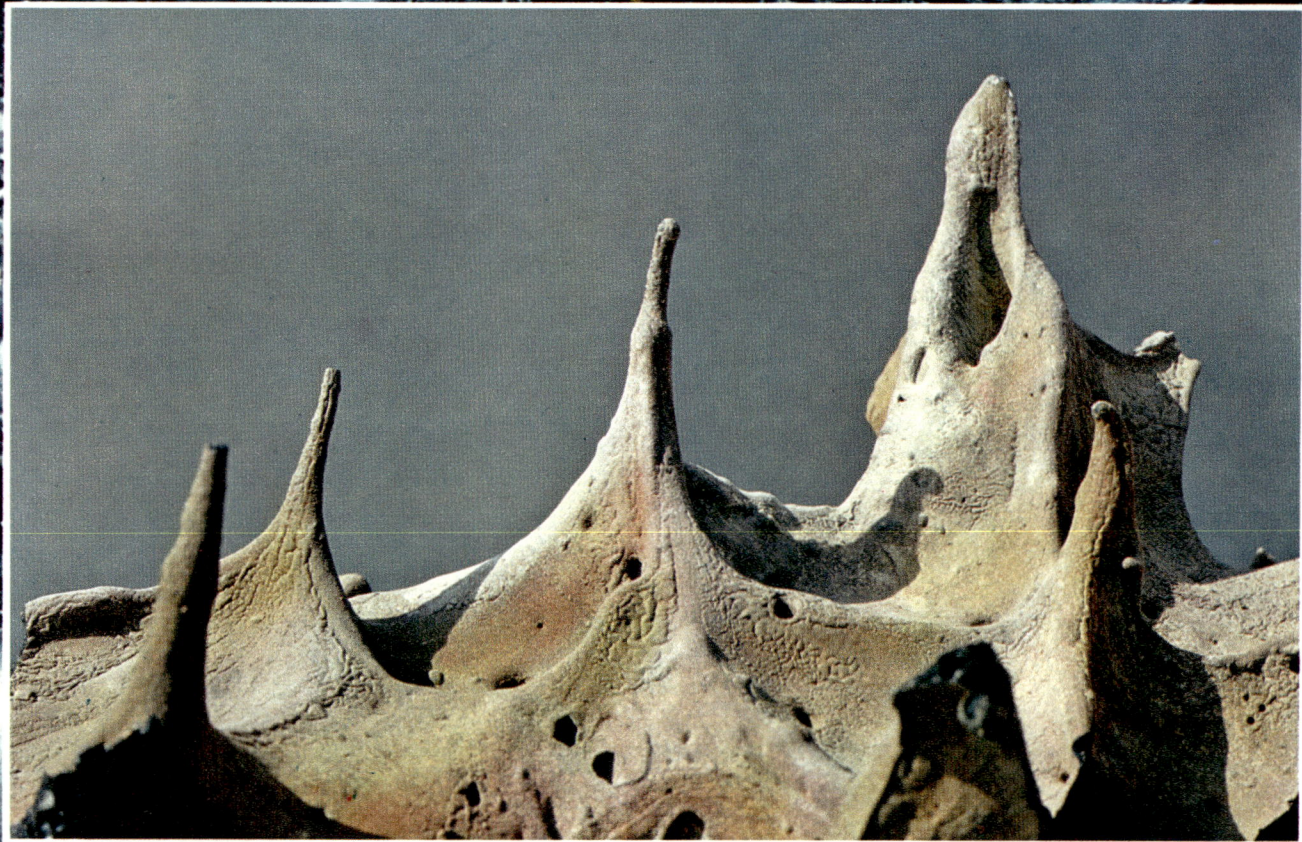

And then,
the most humble of plants,
a moss.
And then,
one morning,
the first sound of an insect,
so dry
you would think it
still mineral.
And then,
hope . . .

◀ Pariou seems to have been born yesterday,
so perfect and well preserved is its form; but
it was several thousand years ago that the
earth opened here and that volcanic spin-
dle bombs, ashes, and scoria accumulated
to form the contours of the crater (Massif
Central, France).

At Nyamuragira, in the uppermost caldera,
entrail and slab lavas, and the sulfurous
fumaroles, recall the activity of the past.
Already lichens and mosses cover the basalt
(Virunga Chain, Zaire).

An impressive flow of obsidian cuts across ▶▶
the valley of Landmannalaugar; its dark
color, spotted with mosses and lichens,
contrasts with the light acid domes in the
background (Iceland).

Then the plants
quicken their desire
to be the first to taste
the mineral juices
of the prime material.
Confusing image
of life
reborn
over the remains
of the consumed god.

◄◄ The cold climate of Iceland and the violence of
the wind retard the formation of a fertile volcanic
soil; however, after hundreds of years, plants
appear in the lapilli, small particles of lava
emitted at the time of the explosion of Hverfjall,
2,500 years ago (Iceland).

These magnificent ropy lavas flowed, fluid, at the
time of an eruption of Nyamuragira, less than a
hundred years ago; they are already covered
with vegetation (Virunga Chain, Zaire).

147

Unmindful
of yesterday's faithlessness,
the houses of men
have resumed
their feverish demands.
All is ready
for all to begin again!
Is the courage of habit
perhaps the finest form
of courage?

◄ In the distance is Mauna Loa, the largest active volcano on earth. The distance from its base at the bottom of the Pacific Ocean to its summit, crowned by a caldera, is almost 10 kilometers—it is thus the highest mountain in the world. The diameter at the base is about 200 kilometers; the slopes are very gradual, with an average inclination of 5°. In the foreground is Mauna Kea, another giant shield volcano (Hawaii).

After six months of activity, the eruption of Eldfell had stopped; only a few fumaroles were left to remind us that there was an active volcano. The Icelanders came back, built new houses, and resumed their lives as fishermen (Vestmannaeyjar, Heimaey, Iceland).

Indonesia has the largest concentration of active, ►► and deadly, volcanoes but also the most fertile and populated soil. The Merapi volcano is the most violent example, with its pelean eruptions with *nuées ardentes*. It is situated where the Indian plate is moving under the Eurasian one, giving rise to the andesitic volcanism of the whole Indonesian arc (Java, Indonesia).

From irrationality to silence,
from the impromptu to the ceremonial,
to you go our prayers,
O Volcano,
brother of serenity and meditation.

Fuji san, or Mount Huzi (3,776 meters high), is situated 100 kilometers
southwest of Tokyo. Majestic in its isolation and perfection, it is visited every
year by tens of thousands of pilgrims. This stratovolcano, whose last
eruption occurred in 1707, is sacred; for the Japanese it is the object of a
genuine cult.

Having become wisdom again,
the mountain has resumed the face of an experienced ancestor.
But does it take such demonstrations
for men to understand the fragility of the frontier
that he treads between existence and death?
Of all the some six hundred active volcanoes
garlanding the weakest areas of the earth's crust,
how many could still contrive Pompeiis?
Strange destiny of little man
who dreams of the stars
while dancing on a film of basalt
as thin as an eggshell.
And under his feet the inner fire,
the fire that seems to burn while being fed by nothing
yet is not loath to consume the fruits of the earth.
Beyond all human morality,
a quickening fury works always toward the future.
Today it is an extinguished volcano,
great beast crouching
in the hollow of the evenings;
it likes to play with twilight and the sea,
inventing for itself necklaces of light,
it is an extinguished volcano.

Klabat is one of the countless volcanoes of the Pacific ring of fire. Its silhouette with steep slopes is that of an andesitic volcano whose viscous lava accumulates around the place of eruption (Celebes, Indonesia).

At the center are the seeds;
at the center is the fire that engenders.
That which sprouts burns. That which burns sprouts.

Gaston Bachelard,
La Psychanalyse du feu

Vesuvius
Engraving by Nicolas de Fer ▶

1. Vieil tronc d'Arbre ou l'on attache les Chevaux.
2. Maniere de Pierre de Couleure de brique
   brulées que le mont a Vomit.
3. Crevasses et Ravines tres Profondes.
4. Pour par venir au Sommet il faut marcher Sans
   routte frayée dans la Cendre y etant arrivé on
   Se couche sur le Ventre d'ou ceux qui n'ont pas
   la temerité de desendre dans les entrailles de ce
   mont Satisfont leurs curiosités.

LE MONT
ou Montagne de
pres de

COUPE DU MONT VESUVE.
A. Fond de la Montagne au
   Niveau de la Mer d'ou s'eleve
   vne autre Montagne de
   cendre dont le Sommet
   est fait en Bassin du quel
   Sort Feu flammes et Soufre et de temps en temps
   des Pierres Brulez d'vne
   grosseur prodigieuses
B. Ruisseaux de soufre

VESUVE
Somma
Naples.

la Marine

les Lacrable

della Fico

MER

MEDITERRANÉE

Par N. de Fer.

Avec Privil. du Roi.

TABLES

Number of Volcanic Eruptions (Above Sea Level) in Historical Times

Pacific Ring of Fire	total 422	Atlantic	total 49
		Jan Mayen	1
		Iceland	21
Kamchatka	20	Azores	13
Kuril'skiye Ostrova (Kuril Islands)	33	Canary Islands	4
Japan	31	Antilles	9
Tibet	5	Tristan da Cunha	1
China Sea	7		
Ryukyu Retto	6	Mediterranean	total 13
Marianas	20	Italy	10
Philippines	15	Greece	3
Andaman Islands	1		
Indonesia	75	Africa	total 23
Melanesia	30	Asia Minor	8
Samoa	4	Ethiopia	4
Tonga–Kermadec	14	Kenya–Tanzania	7
New Zealand	5	Zaire	3
Antarctica	10	Cameroun	1
South America	47		
Central America	42	Indian Ocean	total 4
Galápagos Islands	7	Comores	1
Hawaii	4	Réunion Island	1
North America	7	Heard Island	1
Alaska	39	Saint Paul Island	1

Sizes of Some Large Craters (Calderas)

Volcano and location	Dimensions (in km.)	Volcano and location	Dimensions (in km.)
Toba (Indonesia)	160 x 30	Aso san (Japan)	25 x 17
Buldir (Alaska)	43 x 21	Kawah Idjen (Indonesia)	20 x 16
Valles Mountains (New Mexico)	29 x 25	Bolsena (Italy)	17 x 17

Sizes of Some Large Craters (Calderas)

Volcano and location	Dimensions (in km.)	Volcano and location	Dimensions (in km.)
Emi Koussi (Chad)	15 x 12	Crater Lake (Oregon)	8 x 8
Piton de la Fournaise (Réunion Island)	13 x 9	Rakata (Krakatau) (Indonesia)	7 x 7
Batur (Indonesia)	12 x 9	La Palma (Canary Islands)	7 x 7
Thíra (Santoríni) (Greece)	11 x 7	Trou au Natron (Chad)	8 x 6
Albano (Italy)	11 x 10	Vico (Italy)	7 x 6
Masaya (Nicaragua)	11 x 6	Mauna Loa (Hawaii)	6 x 3
Tengger (Indonesia)	9 x 7	Kilauea (Hawaii)	5 x 5
Deception Island (Antarctica)	9 x 5	Somma–Vesuvius (Italy)	4 x 4

Volume of Volcanic Products Emitted by Some Eruptions

Volcano and location	Year	Volume (in cubic km.)
Thíra (Santoríni) (Greece)	1500 B.C.	70
Tambora (Indonesia)	1815	30
Katmai (Alaska)	1912	30
Rakata (Krakatau) (Indonesia)	1883	18
Vatnaöldur (Iceland)	4600 B.C.	15
Laki (Iceland)	1783	12
Eldgjá (Iceland)	950	9
Bezymyannaya (Kamchatka)	1956	3
Öraefajökull (Iceland)	1362	2
Hekla (Iceland)	1947	1
Shiveluch (Kamchatka)	1964	1

Number of Deaths in Some Eruptions

Volcano and location	Year	Number of deaths
Tambora (Indonesia)	1815	50,000
Rakata (Krakatau) (Indonesia)	1883	35,000
Mount Pelée (Antilles)	1902	28,000
Laki (Iceland)	1783	10,000
Kelud (Indonesia)	1919	5,500
Mount Lamington (New Guinea)	1951	4,000
Vesuvius (Italy)	79	2,000
Soufrière (Antilles)	1902	1,600
Taal (Philippines)	1911	1,300
Merapi (Indonesia)	1931	1,300
Agung (Indonesia)	1963	1,300

Energy Freed by Some Eruptions

Volcano and location	Year	Energy freed in H-bomb equivalents
Thíra (Santoríni) (Greece)	1500 B.C.	4,000,000 H-bombs
Laki (Iceland)	1783	900,000
Tambora (Indonesia)	1815	800,000
Cosigüina (Nicaragua)	1835	400,000
Rakata (Krakatau) (Indonesia)	1883	200,000
Katmai (Alaska)	1912	200,000
Santa María (Guatemala)	1902	50,000
Sakura jima (Japan)	1914	50,000
Etna (Italy)	1669	30,000
Bezymyannaya (Kamchatka)	1956	30,000

Energy Freed by Some Eruptions

Volcano and location	Year	Energy freed in H-bomb equivalents
Mauna Loa (Hawaii)	1859	15,000 H-bombs
Öraefajökull (Iceland)	1362	10,000
Tarawera (New Zealand)	1886	10,000
Hekla (Iceland)	1947	10,000
Shiveluch (Kamchatka)	1964	10,000
Asama yama (Japan)	1783	9,000
Kilauea (Hawaii)	1840	8,000
Fuji san (Japan)	1707	7,000
Vesuvius (Italy)	1906	2,000
Surtsey (Iceland)	1963	2,000

Volcanic Eruptions in 1973

January 9   Paluweh (Indonesia). Large explosion, fall of ashes; the plantations of the island destroyed.

January 13   Pacaya (Guatemala). Emission of large quantities of lava, explosions of ashes.

January 23   Eldfell (Iceland). Opening of a fissure 1,800 meters in length, emissions of fountains, streams of lava; destruction of part of the town of Vestmannaeyjar.

January   Rakata (Krakatau) (Indonesia). Emission of ashes with each explosion.

February 1   Asama yama (Japan). Very large explosion, the cloud of ashes rising to a height of 4,500 meters.

February 22   Fuego (Guatemala). Large explosions of ashes, with clouds rising to 12,000 meters.

April 18   Long Island (New Guinea). A cone formed in the crater lake of the volcano.

May 5   Kilauea (Hawaii). A fissure opened, fountains of lava spurted, a lake of lava formed and overflowed; the lava streams burned the forests.

May 22   Akutan (Alaska). Emission of ashes.

May 30   Nishino shima (Japan). Emergence of a volcanic island in the sea.

June   Sakura jima (Japan). More than eighty very strong explosions in 1973, the clouds of ashes rising to 5,000 meters.

July 12   Mount Lagila (New Guinea). Violent explosions, emergence of flow from the crater.

July 14   Tiatia (Kuril'skiye Ostrova) (Kuril Islands). After 160 years of dormancy, the volcano became active; a new crater was formed.

July 19   Near the reef of Curacoa (Samoa). The sea boiled, pumice was emitted.

September 16   Santa María (Guatemala). Explosion of the dome, emission of nuées ardentes and mudflows.

September 25   Unnamed undersea volcano (West Pacific). Activity remained undersea; it was detected by hydrophones.

October 4   Ulawun (New Britain). Emission of lava flows, explosions and nuées ardentes; forests destroyed.

November 12   Pavlof (Alaska). Strong explosion, emission of a lava flow.

November   Simultaneous eruptions of Wolf and Fernandina (Galápagos Islands). Emission of lava, columns of vapor, collapse of caldera.

In addition to these eruptions, at least seven volcanoes were in constant eruption in 1973.

Two lava lakes: Ertaale (Ethiopia)
Nyiragongo (Zaire)

Five explosive volcanoes: Batur and Semeru (Indonesia)
Yasur (New Hebrides)
Stromboli (Italy)
Mount Erebus (Antarctica)

SELECTED BIBLIOGRAPHY

Aubert de la Rüe, E. *L'Homme et les volcans.* Paris: Gallimard, 1958.

Binggeli, V. *Vulkane.* Bern: Haupt, 1965.

Bullard, Fred M. *Volcanoes: In History, in Theory, in Eruption.* Austin: University of Texas Press, 1962.

Civetta, L., P. Gasparini, G. Luongo, and A. Rapolla. *Physical Volcanology.* Amsterdam: Elsevier, 1974.

Galanopoulos, George A., and Edward Bacon. *Atlantis: The Truth Behind the Legend.* Indianapolis: Bobbs-Merrill, 1969.

Green, Jack, and Nicholas M. Short, editors. *Volcanic Landforms and Surface Features: A Photographic Atlas and Glossary.* New York: Springer-Verlag, 1971.

Krafft, Maurice. *Guide des volcans d'Europe.* Neuchâtel: Delachaux, 1974.

——. *Volcans et tremblements de terre.* Paris: Edition des Deux Coqs d'Or, 1974.

Krüger, C. *Vulkane.* Vienna: Schroll, 1970.

Luce, J. V. *Lost Atlantis: New Light on an Old Legend.* New York: McGraw-Hill, 1969.

MacDonald, Gordon A. *Volcanoes.* Englewood Cliffs, New Jersey: Prentice Hall, 1972.

MacDonald, Gordon A., and Agatin T. Abbott. *Volcanoes in the Sea: The Geology of Hawaii.* Honolulu: The University Press of Hawaii, 1970.

Ollier, Cliff. *Volcanoes.* Cambridge, Mass.: M.I.T. Press, 1970.

Orcel, J., and E. Blanquet. *Les Volcans.* Paris: Bourrelier, 1953.

Rittmann, Alfred. *Volcanoes and Their Activity.* New York: Wiley, 1962.

Wilcoxson, Kent. *Chains of Fire: The Story of Volcanoes.* Philadelphia: Chilton, 1966.

◄ Krakatau in Activity
Engraving by Weber

LIST OF COLORPLATES

PHOTOGRAPH CREDITS

The diagrams were prepared by Maurice Krafft.

The photographs on the following pages are by the members of the Vulcain Team. Jean-Jacques Bacquet: 104–105; Roland Benard: 94; Jacques Durieux: 86–93, 98–99; Katia Krafft: 61–63, 66–69, 71–75, 78–79, 82–83, 100–104, 106–107, 109, 112–17, 120–21, 124, 127–35, 140–47, 150–53, 156; Maurice Krafft: 64–65, 80–81, 84–85, 96–97, 110–11, 118–19, 122–23, 136–39.

The photographs on the following pages are by courtesy of B. Glinn-Magnum: 154-55; Solar Film: 76-77; Donald A. Swanson (Hawaiian Volcano Observatory; U.S. Geological Survey): 148-49.

The engravings are from the album *Le Tour du monde*, published by Editions Planète. Cliché Giraudon, page 23.

Printing completed
August 30, 1975
on the presses of
Draeger